*Dedicated to my son, a sunny boy
who had to grow up so early.*

Chapter I

Fundamentals and Trends of the New Being

> "*I can't describe it: it's like Nausea, only with the opposite sign, in a word, my adventure begins, and when I ask myself where I got it from, I understand what's going on: I FEEL LIKE MYSELF AND I FEEL LIKE I'M HERE; I cut through the darkness and I'm happy, like the hero of a novel.*"
>
> Jean Paul Sartre
> "*Nausea*"

The Age of Pisces and the Age of Aquarius

The Age of Pisces and the Age of Aquarius are not about astrology, magical signs, sacred stones, star caps, prayers[1], or similar practices. While some believe in their efficacy, I am firmly convinced that they are merely placebos, lacking serious influence beyond the individuals who engage in such practices. It is important to clarify that there may be a certain effect, but it is more related to self-hypnosis.

The constellations under which we were born have not remained the same over the billions of years since their light reached us. These photons, which traveled at the speed of 300,000 km/s from the depths of space, created electrical signals in our optic nerves and were processed by our brains through intricate networks of axons and dendrites[2]. However, it is crucial to acknowledge that the constellations no longer appear as they did when their light was emitted.

There exist other objective factors that exert a more significant influence on our lives, as well as the state of our planet. These factors go beyond personal astrological beliefs and are rooted in observable phenomena.

---

[1] Except, by the way, the Aramaic formula "AbraKadabra" ('A Bara' — "*I'm creating*", 'ke Dab´ra' — "*as I speak*") which is a simple way of creating reality through manifestation.

[2] *Axon* — the process is transmitting, through it the impulse goes from the cell body to another neuron. *Dendrites* — receiving processes, they collect impulses from other neurons and transmit them to the body. (In fact, some dendrites conduct the signal in two directions, to and from the body of the neuron.)

Gravity

It is about gravity. Gravity plays a significant role in shaping the unique conditions we experience at the moment of our birth, and our newborn brains perceive these conditions as the fundamental aspects of our environment. Over time, people have observed long-standing behavioral patterns among those born during specific times of the year, leading to the development of general formulas that apply to a majority of individuals.

The Chinese practice of Feng Shui[3], which involves organizing physical spaces, operates on a similar principle. During my time in Hong Kong, I had the opportunity to learn from the son of an experienced Feng Shui master. He shared an example that vividly illustrates the efficacy of this practice and its relevance to our next steps. In Feng Shui, there is a rule that advises against placing the entrance of a house on the north side, as a violation of this rule is believed to bring about failures in family life. This principle holds true in China due to a specific set of circumstances: during winter, a strong gusty wind blows from the Gobi Desert for two months, and when the front door is opened, the wind rushes into the house, instantly cooling it. As children in China often spend time on the floor, they are susceptible to catching a cold. The wife, burdened with caring for sick children, lacks sleep and becomes irritable. The husband seeks solace outside the family, leading to the destruction of the household. This example demonstrates a statistical study based on thousands of years of observation, revealing the

---

[3] *Feng Shui* or *Feng Shui* (Chinese trad. 風水, control. 风水, pinyin fēngshuǐ — literally "*Wind & Water*") is the Taoist practice of symbolic development (organization) of space. With the help of Feng Shui, you can supposedly choose the "best" place to build a house or burial. The stated purpose of Feng Shui is to search for favorable flows of qi energy and use them for the benefit of man.

probability distribution of standard deviation within a Gaussian curve[4].

My journey into understanding these concepts began during my time at the German business school in Vallendar, near Koblenz, where I studied mathematical statistics without fully comprehending its purpose. However, the pieces of the puzzle eventually fell into place. I first encountered the connection between these ideas in a speech by Rebbe Mendel Schneerson[5], who highlighted the symbiosis between the Eastern European soul and the practicality of American culture as a key factor in the connection between the two[6].

---

[4] In mathematics, a Gaussian function, often simply referred to as a Gaussian, is a function of the form $f(x) = ae^{-\frac{(x-b)^2}{2c^2}}$ for arbitrary real constants a, b and non zero c. It is named after the mathematician Carl Friedrich Gauss. The graph of a Gaussian is a characteristic symmetric "*bell curve*" shape. The parameter a is the height of the curve's peak, b is the position of the center of the peak and c (the standard deviation, sometimes called the Gaussian RMS width) controls the width of the "bell".

[5] *Rabbi Menachem Mendl Schneerson* (in lifetime publications and traditionally Hebrew. מנחם מענדל שניאורסאהן Eng. *Menachem Mendel Schneerson*; April 18, 1902, Nikolaev — June 12, 1994, New York), more often referred to as Lubávić(e) Rabbi (Yiddish) דער ליובאַוויטשער רבי, der lubávicher Rébe, Heb. הרבי מליובאוויטש ha Rábi mi Lubá Vitch was the 7th and last Chabad Rebbe. One of the Jewish spiritual leaders of the XX century. He is considered the messiah by some of his followers, some of whom do not recognize his physical death. His grave in New York has become a place of mass pilgrimage.

[6] From Mikhail Epstein's speech at the Liberty Award at 2000: "When I think of Russian American, I imagine an image of intellectual and emotional breadth that could combine the analytical subtlety and practicality of the American mind and the synthetic inclinations, the mystical giftedness of the Russian soul. To combine the Russian culture of pensive melancholy, heartache, bright sadness, and the American culture of courageous optimism, active participation and compassion, faith in oneself and in others..."

5

In a broader sense, the essence lies in the divine union of underlying causes — the desires of the heart — and the American culture of courageous optimism, self-belief, and ceaseless action, often taking precedence over introspective contemplation. This book will delve into the themes of courage and optimism, emphasizing their highest form: valor, which often remains concealed from others due to circumstances, serving as essential components of the path of the ronin.

However, let's return to gravity. Imagine a world where everything is intersected by glowing dotted lines, emitting a faint greenish light. These lines form a three-dimensional mesh, enveloping physical objects and creating elongated shapes where they stretch. In a state of deep meditation and fasting, try holding your hand in front of you. You'll observe how the movement of your hand stretches the space woven by these cells. If you reach a certain level, you'll perceive the changes in time caused by gravity. In my case, a recent cut on my arm mysteriously vanished. Whether it truly disappeared or simply appeared to under the influence of my meditation and fasting is irrelevant. What matters is that this vision aligns with the scientific knowledge of our most advanced researchers, which I'll delve into later in this book.

This phenomenon is known as time expansion. To simplify, imagine a clock placed in a region with a stronger gravitational field. It would tick slower due to the minimal effect of gravity on light, which takes less time to reach a reflective surface, thus slowing down the clock. While this explanation may seem absurd, it aligns with the prevailing physics of our time. It may sound outlandish, but it works.

My hero and unwavering ronin of all eras (the concept of a ronin will be discussed in this book), Stephen Hawking[7], was one of the first contemporary thinkers to

---

[7] Professor Natural Philosophy at Hopkins University and Professor in the Department of Fractals at the Santa Fe Institute, California.

explore the connection between gravity and time within the context of a singularity, a topic I'll cover later. I've selected a concise and popular excerpt from Sean Carroll's work, "The Scientific Legacy of Stephen Hawking":

During his time at Cambridge in the 1960s, Hawking embarked on his doctoral research, driven by an intense curiosity about the universe's origin and ultimate destiny. General Relativity, Einstein's theory encompassing space, time, and gravity, provided the ideal framework for exploring this puzzle. According to general relativity, gravity emerges from the curvature of space–time. By understanding how matter and energy create this curvature, we can predict the universe's evolution. This period can be referred to as Hawking's "classical" phase, distinguishing it from his subsequent investigations in quantum field theory and quantum gravity.

Around the same period, Roger Penrose from Oxford presented compelling evidence: according to general relativity, under a wide range of conditions, space and time would collapse inward, forming a singularity. If gravity represents the curvature of space–time, a singularity denotes a moment when this curvature becomes infinitely immense. Penrose's theorem demonstrated that singularities were not mere curiosities; they played a crucial role in general relativity.

Penrose's insight also extended to black holes, regions in space–time where gravity is so powerful that not even light can escape. Deep within a black hole lies a singularity in its future. Hawking took Penrose's idea and reversed it, applying it to the past of the universe. He showed that, under the same general conditions, space would have emerged from a singularity: the Big Bang.

Present-day cosmologists discuss both the Big Bang model, a highly successful theory describing the universe's expansion over billions of years, and the Big

Bang singularity[8], which remains an enigmatic concept awaiting further understanding.

Hawking then focused his attention on black holes. Penrose's calculations revealed another intriguing outcome: energy could be extracted from a black hole by depleting its rotation until it ceased. Hawking demonstrated that, although energy extraction was plausible, the area of the event horizon surrounding the black hole would always increase in any physical process.

This "area theorem" held significance not only within its own realm but also in relation to an entirely different branch of physics: thermodynamics, the study of heat transfer. Thermodynamics adheres to a set of well-known principles. For instance, the first law states that energy is conserved, while the second law declares that entropy, a measure of the universe's disorder, never decreases in a closed system. Collaborating with James Bardeen and Brandon Carter, Hawking proposed a set of "black hole mechanics" laws akin to thermodynamics. Similar to thermodynamics, the first law of black hole mechanics ensures energy conservation, while the second law, known as the Hawking Area Theorem, asserts that the event horizon's area never diminishes. In essence, the area of a

---

[8] *Singularity* (from Latin. *singularis* "the only one, special")
- *Singularity in philosophy* is the singularity of a being, event, phenomenon.
- *A mathematical singularity* is a point at which a mathematical function tends to infinity or has any other irregularities of behavior.
- *Gravitational singularity* (singularity of space–time) is a region of space–time through which it is impossible to smoothly continue the geodesic line included in it.
- *The cosmological singularity* is the state of the universe at the initial moment of the Big Bang, characterized by infinite density and temperature of matter.
- *The technological singularity* is a hypothetical moment, after which technological progress will become so rapid and complex that it will be incomprehensible.

black hole's event horizon behaves much like the entropy of a thermodynamic system — it increases over time.

There are compelling studies on black holes where this principle manifests, impacting matters of life and death. Arguing against this would prove futile.

With this digression, I aimed to emphasize the essential notion that gravity not only interacts with but directly shapes space and time.

PRECESSION[9] OF THE EARTH

Precession of the Earth, also known as axial precession or simply precession, refers to the slow and continuous change in the orientation of the Earth's axis of rotation over a period of time. It is caused by various factors, including the gravitational forces exerted by the Sun, the Moon, and other celestial bodies.

The Earth's rotation is not perfectly aligned with its axis of angular momentum. The axis of rotation is tilted at an angle of approximately 23.5 degrees relative to the plane of its orbit around the Sun. This tilt is responsible for the changing seasons on Earth.

However, the gravitational forces exerted by the Sun and the Moon on the equatorial bulge of the Earth's shape cause a torque that influences the orientation of the Earth's axis. Over a period of about 26,000 years, this torque causes the Earth's axis to trace out a circular path on the celestial sphere. This phenomenon is known as precession.

The precession of the Earth has several effects. One of the most significant effects is the changing pole star. Currently, the North Star, or Polaris, is close to the North Celestial Pole, making it a useful reference point for navigation. However, due to precession, the North

---

[9] *Precession* (from Latin. *praecessio* — movement ahead) is a phenomenon in which the axis of rotation of the body changes its direction
   in space.

Celestial Pole slowly moves over time, and in about 13,000 years, Vega will become the North Star.

Precession also affects the timing of the seasons. The tilt of the Earth's axis relative to its orbit remains constant, but the orientation of the axis changes over time. As a result, the timing of the seasons slowly shifts over thousands of years.

It's important to note that precession is a slow and gradual process that occurs over long timescales. Its effects are not easily noticeable in day-to-day life or even within a single human lifetime. However, over centuries and millennia, precession has significant implications for celestial navigation, astronomy, and the long-term climate patterns on Earth.

The precession of the Earth refers to the gradual change in the orientation of the Earth's axis of rotation. This deviation occurs during the Earth's rotation and is caused by various factors, including the gravitational forces exerted by the Sun, the Moon, and other celestial bodies.

The entire cycle of the Earth's precession takes a little less than 26,000 years and is divided into approximately 12 equal time cycles, each spanning around 2,000 to 2,160 years. Based on the constellation that the Earth aligns with during the vernal equinox, different eras are designated, such as the Eras of Leo, Taurus, Aquarius, Pisces, etc. After the completion of one cycle of eras, the sequence begins again in the opposite direction.

While the analogy of a spinning top can help visualize the basic principles of precession, it is important to note that the dynamics of a spinning top do not fully capture the complexity of the Earth's precession. Additionally, the connection between precession and astrological eras is not yet scientifically supported and is based on esoteric beliefs rather than established scientific understanding.

Understanding the impact of gravity on Earth is of great interest because gravity is the dominant force that governs the behavior of all objects and inhabitants on our planet. Gravity plays a crucial role in shaping Earth's systems and processes. Here are a few reasons why the study of gravity's impact on Earth is significant:

1. *Tides*[10]: Gravity, particularly the gravitational interaction between the Earth, the Moon, and the Sun, is responsible for the phenomenon of tides in the oceans. The gravitational forces exerted by these celestial bodies cause the water on Earth's surface to bulge, resulting in the rising and falling of sea levels. Tides have important implications for navigation, coastal ecosystems, and energy generation from tidal power.

2. *Orbital Dynamics*: Gravity is vital for understanding the motion of objects in space, including artificial satellites and the Moon's orbit around the Earth. Precise knowledge of gravitational forces is essential for space exploration, satellite communications, and accurate celestial navigation.

3. *Planetary Motion*: Gravity is central to our understanding of the Earth's orbit around the Sun and the motion of other planets in the solar system. The gravitational pull of the Sun keeps the Earth in its elliptical orbit, and the interactions between celestial bodies influence the stability and dynamics of the solar system.

---

[10] *Tides* — periodic fluctuations in the level of the ocean or sea, which are the result of the action of the tidal forces of the Moon and the Sun, but the tidal force of the Moon is 2.17 times greater than the tidal force of the Sun, so the characteristics of the tide mainly depend on the relative position of the Moon and the Earth.

Tides cause changes in sea level elevation, as well as periodic currents known as tide currents, making tide prediction important for coastal navigation.

4. *Geodesy and Earth's Shape*: The study of gravity helps us determine the shape and size of the Earth. Variations in gravity across different regions provide insights into the distribution of mass within the planet. Measurements of gravity are used in geodesy to precisely define the Earth's geoid (a hypothetical surface that represents mean sea level) and to map the Earth's gravitational field.

5. *Earthquake and Volcano Monitoring*: Gravity measurements can be used to monitor changes in the Earth's crust, such as subsidence or uplift, which may indicate geological processes like tectonic plate movements, volcanic activity, or underground water storage. This information is crucial for assessing seismic hazards, volcanic eruptions, and land subsidence.

> The close relationship between the ocean and the atmosphere and the influence of gravity on climate are indeed important factors to consider. The ocean plays a significant role in shaping weather and climate patterns, and understanding its behavior is crucial for predicting and studying climate dynamics. Here are some key points to emphasize:

1. *Solar Energy Absorption*: The ocean absorbs a substantial portion of the solar energy that reaches the Earth's surface. Due to the uneven distribution of solar radiation, with equatorial zones receiving more energy than circumpolar regions, the ocean acts as a reservoir that accumulates and transports heat energy across the planet[11].

---

[11] The coefficient of absorption of solar energy in the ocean and land, of course, is different, but the following facts can give a sense of the overall picture: 71% of the Earth's surface is water, and the remaining 29% consists of continents and islands. At the same time, 96.5% of all the Earth's water is contained in the oceans in the form of salt

2. *Ocean Currents*: Horizontal and vertical ocean currents play a crucial role in distributing heat energy. Prominent currents like the Gulf Stream[12] transport heat over long distances, transferring it to the atmosphere. The ocean's slower warming and heat transfer compared to land result in more moderate weather conditions along coastal areas.

3. *Energy Distribution and Weather Systems*: The redistribution of energy by ocean currents contributes to the formation of weather systems, including cyclones and storms. Changes in the angle of gravity's application on the Earth's surface can significantly affect climatic factors due to alterations in energy distribution.

4. *Ocean's Role in Climate Regulation*: The distribution of land and water on Earth influences climate. The combined effects of oceanic and atmospheric circulation, driven in part by gravity, account for a significant portion (up to 70%) of the total influence on climatic factors.

5. *Ocean's Heat Storage*: The ocean plays a crucial role in mitigating climate change by absorbing excess heat resulting from anthropogenic carbon dioxide emissions. Approximately 90% of the accumulated excess heat is stored in the ocean, with only a small

---

water, and the remaining 3.5% is freshwater lakes and frozen water enclosed in glaciers and polar ice caps. At the same time, 69% of this part of fresh water, almost all of it is in the form of ice.

[12] *Gulfstream* (from the English. *gulf stream*. The Gulf Current is a warm sea current in the Atlantic Ocean. In a narrow sense, the Gulf Stream is called the current along the east coast of North America from the Florida Strait to the Newfoundland Bank (as it is, in particular, marked on geographical maps). In a broad sense, the Gulf Stream is often referred to as a system of warm currents in the North Atlantic Ocean from Florida to the Scandinavian Peninsula, Svalbard, the Barents Sea and the Arctic Ocean.

fraction (about 2,3%) warming the atmosphere. The rest contributes to ice melting and warms the land.

6. *Gravitational Influence*: The state of the world's oceans and, consequently, the climate on Earth, is directly related to the gravitational forces exerted by the Sun and the Moon. The Moon's influence is approximately two-thirds, while the Sun's influence constitutes about one-third. This division is associated with the distance between these celestial bodies and the Earth.

Considering the complex interplay between the ocean, atmosphere, and gravity, scientists strive to study and model these interactions to enhance climate predictions and improve our understanding of the Earth's climate system.

The discovery of a book from 1969[13] revealed that precession was discussed as a main cause of social change in ancient history. The authors argue that precession held significant importance in ancient cultures, being encoded in their myths, symbols, and religious beliefs. They viewed precession as a sacred and mystical cycle that reflected the nature of time and the universe.

According to the authors, ancient myths and symbols, including constellations, were used to represent the precessional cycle.

The authors propose that precession played a crucial role in the development of astronomy and astrology in ancient cultures. Ancient astronomers utilized precession to track the movements of stars and planets, which aided in creating calendars and predicting celestial events.

---

[13] *"Hamlet's Mill"* is a book by Giorgio de Santillan and Herta von Dechend that explores the idea that ancient myths and symbols reflect knowledge about astronomical cycles, especially the precession of the equinoxes.

In addition, the authors suggest that precession influenced the construction of ancient monuments and temples. Notably, structures like Stonehenge and the Great Pyramid of Giza were associated with astronomical events related to precession.

The ancient knowledge of precession was passed down through generations by astronomers and priests, shaping human culture and civilization. The authors speculate that precession may have played a key role in the development of religions and spiritual beliefs of ancient times, influencing how ancient cultures perceived the universe and their place within it.

Overall, the theory presented in the book "Hamlet's Mill" emphasizes the fundamental nature of precession in ancient mythology and culture, asserting its profound impact on human history. The authors contend that precession played a role in the development of astronomy, astrology, religion, and architecture, influencing the worldview of ancient cultures. Despite being relatively unknown, this theory has had a significant influence on the study of ancient myths and symbols, remaining a subject of discussion in academic circles, for example the authors believed that the myth of the constellation Orion[14], for example, represented the precession cycle, with the rising and setting of Orion marking the beginning and end of the cycle.

---

[14] *Orión* (ancient Greek Ὠρίων). In ancient Greek mythology, he was a famous hunter who was distinguished by his extraordinary beauty and such growth that he was sometimes called a giant. Once Zeus, Hermes and Poseidon (or Ares), and according to Michael Mayer (XVII century) — PhoebusApollo, Vulcan and Hermes, visited the inhabitant of Thebes Hyriaeus. When he, after sacrificing a bull and treating the gods, began to complain of childlessness, the guests demanded the skin of the victim. When the owner brought the skin, they filled it with urine and ordered it to be buried in the ground. Nine months later, Orion emerged from it.

PREVIOUS ERAS

We are currently in the middle of the astronomical cycle of eras, with the next one beginning approximately 13,000 years from now. Let's briefly explore the characteristics of the previous eras:

Age of Leo
(approximately 10,800 BC to 8,600 BC)

This era was characterized by a growing sense of individuality and a desire to become leaders and gain power. In primitively organized societies, there was a separation of ruling classes, and a direct connection with nature, the universe, higher worlds, and God. The approach towards death was balanced and calm.

Age of Cancer
(approximately 8,600 BC to 6,500 BC):

During this era, there was a shift towards forming households, constructing houses, and settling into family-oriented communities. Agriculture and farming practices emerged, and Mother Nature, femininity, and women were deified. The initial religious forms revolved around animal cults.

Age of Gemini
(approximately 6,500 BC to 4,000 BC)

Settlements started evolving into semblances of modern cities during this era. Concepts of trade, road communications, and connections began to emerge. Writing ceased to be a privilege limited to the upper classes and gradually became more commonplace. Arts, skills, crafts, and logic took precedence, replacing intuition.

Taurus Era
(approximately 4,000 BC to 2,000 BC)

This era witnessed advancements in irrigation techniques and agriculture. Massive structures reflected people's desire for tranquility and comfort, symbolized by

Taurus. The focus shifted towards studying the material aspects of well-being, and the connection with subtle energies diminished. The cult of women as providers of comfort and security prevailed.

Age of Aries
(approximately 2,000 BC to 0 AD)

The Age of Aries saw the peak of early Hellenic culture, as well as Jewish and Persian civilizations. The cult of male power, exemplified by Sparta, emerged. The role of women in society was suppressed and diminished. Strength and the cult of the strong became prominent, and a "tit-for-tat" mentality prevailed. The development of metallurgy also characterized this era.

Age of Pisces
(approximately 0 AD to 2,000 AD)

Christianity dominated this era, marked symbolically by the sacrificial slaughter of the lamb, signifying the end of the Age of Aries. Compassion, the cult of sacred suffering, and sacrifice for others were prominent. The emergence of high culture and art fueled by mystical inspiration occurred. Acts of charity, forgiveness, empathy, and, conversely, extreme manifestations of cruelty and violence were prevalent. The era involved self-exploration of faith without tangible proof of the divine principle. Victimization and martyrdom, despite their difficulties, led to a softening of character and the practice of compassion. Connections with higher worlds were developed, though the process was painful yet necessary.

Age of Aquarius
(approximately 2,000 AD to 4,000 AD)

The Age of Aquarius is characterized by grounded knowledge and science as key factors. This era began approximately 200 years ago with industrialization and mechanization, gradually progressing towards a high-tech civilization. The pinnacle of the Age of Aquarius is defined by the development of science, which liberates humanity to nurture individual talents and personal individuality, moving away from the struggle for survival. However, there are also lower aspects of this era that require lyrical reflection for a deeper understanding.

In a lyrical digression, a Kabbalistic parable is mentioned about a man seeking wisdom from Abraham[15]. Through fasting and spiritual purification, the man achieves moments of elevated consciousness, akin to a prophet, but cannot reach the same.

To attain wisdom from Abraham, the Kabbalist embarked on a journey to the desert, engaging in a 40-day fast, practically depriving himself of sustenance. Through this intense spiritual discipline, he aimed to elevate his spirit to the level of a prophet. Despite his diligent efforts and significant purification of his *klipot*[16] (spiritual impurities), he reached remarkable heights but fell short of reaching the same spiritual level as the deceased prophet.

Undeterred, the Kabbalist chose to undergo another 40-day fast, pushing himself to the brink of life and death. In this state of extreme physical and spiritual deprivation, he managed to briefly ascend to the level of a prophet, transcending his usual limitations. However, he was aware that these elevated states are fleeting, and one is inevitably brought back to their ordinary state. With this understanding, the man seized the opportunity to ask the prophet for the most important piece of wisdom.

The prophet nodded in response and uttered a seemingly peculiar answer: "Where there is more darkness, there is

---

[15] *Abraham* (Eng. *Avraham Avinu* or Aram. (אבינו אברהם), "*our Father Abraham*", meaning both the biological progenitor of the Jews and the spiritual father of Judaism.

[16] *Clipót* (Eng. *Klipot*) Borrowed from Aram. multiple. *Klipot* (*klipa*, unity.) translated from Aramaic as "peel", "husk" — means metaphysical barriers between us and the Light of the Creator, which we have created ourselves with our limited selfish actions. They do not allow us to constantly feel happy and confident in the future.

more light." This enigmatic response left the man puzzled, but perhaps with a renewed perspective. After enduring 80 days of fasting and the strange encounter, he likely indulged in a full breakfast for the first time in a long while.

Despite the perplexity of the situation, it prompts us to contemplate deeper questions. Where can we find greater benefits? Is it in established territories with predictable rules or in uncharted territories that require risk-taking? Which relationships are safer? Are they the ones experienced in a utopian paradise or the ones where partners willingly put their lives on the line to protect each other? This conveys the profound law of existence.

Adding to this well known concept, I would like to share a personal parable[17]: In a near death experience back in the time, I have managed to grasp a pen and a notebook with fingers that refused to cooperate. Despite the circumstances of the accident and those responsible for it remaining unaddressed, I crookedly wrote a single word, "Heart!" with an exclamation mark, emphasizing it three times. As consciousness slipped away, they pondered why, across various traditions, the organ responsible for pumping blood is attributed with the mystical ability to guide its owner in making the most accurate decisions.

CAN THE HEART FEEL?

Traditionally, the heart has been regarded as the source of emotions, desires, and wisdom. However, from a medical science perspective, the functional role of the heart was limited to that of a "circulatory pump" for blood.

In 1991, Dr. Armor[18] made a significant discovery that the heart possesses its own *"little brain"* or *"internal*

---

[17] A short simple story illustrating a moral or spiritual truth.

[18] Intrinsic Cardiac Neurons J. ANDREW ARMOUR M.D., Ph.D.
First published: August 1991
https://doi.org/10.1111/j.1540-8167.1991.tb01330.x

cardiac nervous system". This intricate neural network, consisting of approximately 40,000 neurons, bears structural similarity to neurons found in the brain. Known as the intrinsic cardiac nervous system (ICNS), this "heart brain" is capable of reflex control over heart function, even when isolated from the central and intrathoracic levels of the extracardiac ganglia and the central nervous system. Therefore, the heart is not merely a "pump" but an organ with its own autonomous nervous system, capable of functioning independently of the brain.

Research has revealed that the heart's nervous system can modulate areas in the brain associated with the cognitive and emotional aspects of pain. Thus, the heart likely plays a significant role in regulating pain. Furthermore, the neurons in the heart possess both short-term and long-term memory, and the signals they send to the brain can influence emotional experiences. Remarkably, the heart transmits more information to the brain than the brain sends to the heart.

The signals emanating from the heart and received by the brain can impact perception, including the perception of pain, as well as the assimilation of emotional experiences and other higher cognitive processes. For instance, a coherent heart rate has been shown to aid the brain in solving creative problems. It is important to consider that the heart lacks direct access to the senses, rendering it less susceptible to deception by negative experiences, external influences, upbringing, or public opinion. Therefore, decisions guided by the heart are likely to be the most authentic and appropriate for an individual. Listening to the heart is essential.

These contemplations and musings invite further exploration and discussion into the deeper meanings and significance of the heart as an instrument of intuitive guidance in human existence.

LOW LEVEL OF AQUARIUS

The low level of the Age of Aquarius signifies the disconnect between the rapidly evolving mind and the heart. A valuable piece of advice I received at a young age, from a book by a French writer, emphasized the importance of listening to the heart for true happiness. It took me 42 years to fully comprehend the meaning behind these words. Antoine de Saint Exupery[19] eloquently wrote, "Only the heart is vigilant. You can't see the most important thing with your eyes."

Practically speaking, this implies recognizing the destructive misuse of reason, such as aggressive artificial intelligence promoting unnecessary products for profit, enslavement through mindless consumerism, the instigation of wars for territorial and power gains, and the prevalence of ideologies like fascism, communism, corporatism, and religious fanaticism.

However, amidst this confrontation with a lower level of consciousness, I find solace in the support provided by Nature itself. It offers us boundless resources of geothermal, solar, tidal, and wind energy. It unveils the potential for teleportation, telepathy, clairvoyance, and the profound experiences of friendship, sympathy, love, and above all, compassion.

Embracing these qualities and recognizing the significance of the heart can lead us towards a higher level of consciousness and a more harmonious existence in the Age of Aquarius.

Thus, the traditional blood ties will gradually diminish and transform into a brotherhood based on shared values and beliefs. This shift will have a crucial implication for the world: globalization, which receded in the face of increasing military conflicts in the previous era, will resurge. Developed nations will extend assistance to

---

[19] Antoine de Saint Exupery. "*The Little Prince.*"

smaller and developing countries, ultimately leading to a unified global governance. This process will entail a confrontation between the East and the West, which will resolve within a relatively short historical time-frame of five hundred years. However, the overarching trends can already be discerned. The fundamental issue of globalization will be addressed through equal and fair access to resources.

The themes explored by Machiavelli in "The Prince" and exploited by generations of fascists belong to the past. The absurdity and futility of such approaches are becoming apparent to only a few, but it is important to observe the established trends.

During the writing of this book, I had the opportunity to attend a theater production of "Fedora," a tragedy based on the works of Seneca. The actors skillfully and splendidly performed this soul-stirring tragedy. However, it no longer captivates the soul as it once did; it seems nonsensical. In the Section Eight high-rise building for the poor in the Bronx, New York, similar tragedies unfold twice a week, yet they no longer hold the same significance. The global trends have changed. They have not merely shifted; they have already transformed, and the results are now evident.

The strategy of "Divide and Conquer" is rapidly fading into oblivion. In fact, it has already become obsolete[20].

Dictators around the world observed with dread as the Internet, despite its imperfections as a mass communication medium, gained significant strength. Looking ahead, it is worth noting that even more advanced forms of universal communication await us. For instance,

---

[20] I will write in small text about the attack of fascist Russia on the growing democracy of Ukraine — this question seems so small to me in the context of this book precisely because the question of the future of the world is decided by much more powerful forces than the most secret of the powerful societies of the planet — in fact, the question is decided by the planet itself.

telepathy as a means of interpersonal communication or accessing the "world of ideas" may become a reality. This realm, known as Sefira Bina in classical Kabbalistic theory, has been recognized since ancient times.

In our exploration, we find particular interest in Bina[21] and its counterpart, Hochma, as they appear to be opposing forces but are, in fact, inseparable from one another. Hochma represents "Wisdom," embodying a masculine and active principle, while Bina symbolizes "Mind," reflecting a feminine and passive aspect. Together, Hochma and Bina unite to form Daat, which signifies "Knowledge[22]."

In a forthcoming chapter of this book, we will delve into practical applications of the insights derived from these paired Sefirot, enabling individuals to harness their inherent wisdom and mental capacities for personal material benefit.

From my personal perspective, I share the belief that the conclusion of the Age of Aquarius will bring about universal and individual access to the vast information of the Universe, as well as direct communication among people without intermediaries. This shift will lead to a rejection of direct interactions with traditional elites and a transformation of the very concept of elitism. As an intermediate stage, there may be a consolidation of states into three super-countries, akin to the Orwellian[23] notion.

---

[21] *Biná* (Aram. בינה; bīnāh; "Mind"; "Mind"; "Thought") — in the teachings of Kabbalah on the origin of the worlds is the third of the 10 objective emanations (direct rays of divine light) of the universe — the so-called "*Sefirot*" or "*Sephiroth*" (plural from "*Sephirah*"), also "figures" or "spheres", the first radiations of the Divine Essence, which together form the cosmos.

[22] *Daat* (Dr.Heb. דעת; "*Cognition*") — the contrast between subjectivity and objectivity finds its resolution in "cognition".

[23] *George Orwell* (Eng. *George Orwell*, real name Eric Arthur Blair; June 25, 1903 – January 21, 1950) was a British writer, journalist

However, with the advent of advanced communication technologies, the role of elites will gradually evolve into that of intellectual leaders guiding humanity on our planet. Consequently, the function of the state will simplify to primarily administrative responsibilities.

In fact, the future President of planet Earth, considering a run for office in the next 600 years, will carefully contemplate whether they possess the fortitude to devote their life to waste management and resource regeneration. In a shorter timeframe, even within our lifetime, following the projected robotization singularity of 2029–2030, our creative abilities may be sought after by cities that will compensate us for our time or, more accurately, our entire lives, in support of local development, providing us with comfortable living conditions. However, preceding these developments, we must navigate the anticipated confrontation between the Lion and the Dragon, as tradition suggests. It is within these captivating and potentially auspicious times that we, along with our descendants, have the opportunity to live and thrive.

INFLUENCE OF ELEMENTS

The study of the influence of elements is crucial as it provides us with an opportunity to comprehend the context of transitioning into a new era. Without this understanding, the transition itself would become another fragmented chapter in history that we would need to analyze retrospectively. It is of utmost importance for us to know what lies ahead before it occurs, ideally with ample time for adequate preparation. In this endeavor, the elements serve as our guide. As a reminder for those who

---

and literary critic. His works are distinguished by a simple style of presentation, criticism of totalitarianism and support for democratic socialism. Orwell's most famous works are the satirical novel "Animal Farm" (1945) and the novel "1984" (1948).

may have missed the initial pages, the term "elements" refers to the conventional designation for the factors that exert an influence on the archetypes of human behavior, which are in turn influenced by the combination of time, space, and gravity.

While Jung's archetypes[24] mark the starting point of our exploration, they are not the final destination; they merely scratch the surface of the topic at hand. Additionally, there exist fundamental archetypes that shape human behavior. For the sake of convenience and ease of use, I will maintain the terminology of astrology to describe these archetypes, as it proves to be a practical framework. The four elements, each possessing unique properties, align with specific time intervals along the precession axis of planet Earth as follows:

Fire — Aries, Leo, Sagittarius
Earth — Taurus, Virgo, Capricorn
Air — Gemini, Libra, Aquarius
Water — Cancer, Scorpio, Pisces

The elemental dominance undergoes a shift every two hundred years. A few years ago, the dominant element transitioned from earth to air. Let us first examine the characteristics of both elements before exploring the remaining aspects.

From around 1802, the prevailing element was earth, but since the end of December 2020, there has been a change. The world is no longer governed by the earth element;

---

[24] *Carl Gustav Jung* (in German. *Carl Gustav Jung* [ˈkarl ˈgʊstaf ˈjʊŋ]; July 26, 1875 – June 6, 1961) was a Swiss psychiatrist and educator. The scientist identified 12 personality archetypes: "hero" (The Hero), "simple-minded" (The Innocent), "seeker" (The Explorer), "creator" (The Creator), "glorious fellow" (The Regular Guy), "lover" (The Lover), "ruler" (The Ruler), "caring" (The Caregiver), "sage" (The Sage), "rebel" (The Destroyer), "magician" (The Magician), "jester" (The Jester).

instead, the dominant influence is now the element of air. This shift is expected to continue for the next two hundred years. It is important to note that I employ the terms of astrological feng shui for convenience. However, these observations, accumulated over thousands of years, have unveiled the impact of environmental forces, particularly gravity, on societal norms and behavior[25].

Based on my personal hypothesis, the mesh matrix of antimatter undergoes stable deformation under specific gravitational fields, objectively affecting all aspects of biological matter in terms of speed and susceptibility to change. This primarily relates to highly organized brain tissue, leading to fundamentally different functional demands. This influence is most apparent when considering the prism of self-interest or, in other words, the manifestations of the "ego." Continuing along this line of reasoning, we are confronted with practical methods of rectifying the ego when communicating with the universe, the Higher Self, or any transcendent context.

Now, let's briefly explore the characteristics of the earth element. It is associated with a rigid vertical hierarchy, consumerism, and the primacy of material values. However, as we transition into a new era, these phenomena will gradually diminish in importance, giving way to the prominence of intellect, information, and cooperation.

---

[25] This ratio or, say, an index of the direct influence of gravity, is a separate subject of study of science.

This context was first predicted by Albert Einstein in the framework of the Gutenberg project (*Einstein, A. Relativity: the Special and General Theory by Albert Einstein*. I express my gratitude to the Gutenberg Project for access to the original work, but I also want to note that this aspect of gravity was studied separately in tests of general relativity in an attempt to find an experimental basis for the theory of general relativity. Actually, the first three tests were conducted by Einstein himself, who proved the anomalous precession of the perigee of Mercury, and much more importantly, the deflection of light (!) in gravitational fields.

Creativity and communication will become highly valued currencies, and the shift away from material values will be perceived as a natural progression.

Concisely, when considering the land as an element, it is characterized by a rigid vertical hierarchy, consumerism, and a prioritization of material values. However, as we transition into a new era, these aspects will gradually lose significance in favor of intellect, information, and cooperation, which will gain prominence. Creativity and communication will emerge as the most valuable assets, while moving away from a materialistic context will be seen as entirely natural.

Practically speaking, within the Age of Aquarius and the current air period, there is a notable shift towards rational thinking and intellectual development across various levels of perception. This ongoing process is evident in the growing importance placed on mental values compared to religious faith and emotional inclinations prevalent during the preceding Pisces era.

Have you ever pondered why certain historical contexts appear to us as if they possess timeless wisdom, while others seem nonsensical and irrelevant? Let's take the example of a traditional Jewish practice on the evening before Shabbat. During the preparations for kiddush, the faithful engage in reading the Psalm of David — a powerful formula for communion with God that instills tremendous courage and imparts profound wisdom[26].

---

[26] The Lord is my shepherd; I will not lack anything. In the magnificent meadows he will give me rest, he will lead me to calm waters, he will calm my soul; will lead me straight paths for His Name's sake. Even if I pass through the gorge in the darkness of the grave, I will not be afraid of evil, for You are with me. Thy counsel and encouragement will comfort me. Thou shalt set the table before me in full view of my enemies, thou shalt smear my head with oil, and my cup shall be full. May only kindness and love accompany me all the days of my life, and I will be in the temple of the Lord for many years.

However, preceding this, there is a description of an ideal wife who wakes up early, takes care of all household tasks, engages in various trades and activities, tends to the children, and shows meekness towards her husband. It even states that "her lamp will not go out even at night." On the other hand, the husband is advised to sit with the elders at the gate, suggesting a time for contemplation. Yet, there is no mention of the main purpose of this contemplation. This apparent imbalance between the divine wisdom in the Psalm of David and the portrayal of an ideal wife resembling a servant from ancient times raises a question: Why?

The answer lies in the repetition of the elements during the precession cycle. As we know, the precession cycles themselves repeat every 26,000 years, with gradual intermediate cycles punctuated by energetic shifts called singularities. Within these cycles, every 800 years, certain elements, like the description of the ideal wife, may hold significance based on the demands of that time. However, over the span of 24–26 thousand years, this description again gains the profound meaning of divine wisdom.

Reflecting upon this notion, I was overwhelmed by the profound realization that I had touched upon the fabric of universal determinism[27].

To put it simply, the elements repeat themselves three times within the precession cycle, and at the time of writing this book, we are in the early stages of the "air" element. It is essential to examine each of the three air signs — Gemini, Libra, and Aquarius — in sequence to grasp their distinct characteristics, which primarily revolve around communication.

Gemini embodies its essence through everyday communication and the various forms it takes in daily

---

[27] Looking beyond our lives, I want to rejoice in the way I/ We, the mind that stands behind us, are magnificent.

life. Libra, on the other hand, seeks broader and more impersonal communication, often addressing larger groups of people within social contexts of general significance. An illustrative example would be a teacher in a classroom setting.

Aquarius, the third air sign, expresses its nature in the most impersonal manner possible. It goes beyond addressing a specific group and instead delves into abstract concepts that may not have immediate practical applications. However, this exploration enables the establishment of universal laws that have global relevance. A scientist engaged in theoretical physics and popularizing scientific knowledge through television programs and the internet is an exemplar of an Aquarius-like individual. They tackle communication challenges with a global scope.

Undoubtedly, such developments in communication will bring about various changes in human morality. For instance, consider the transformations that copyright laws may undergo in response to shifting contextual demands and mankind's increasing interest in dissemination. This interest will increasingly focus on intellectual rather than sensory information, shaping the way we understand and interact with knowledge in a globalized world[28].

The shift from the earth element to the air element is clearly evident in the rise of companies whose mission aims to benefit humanity on a global scale, as well as the popularity of movies, shows, and other forms of entertainment that explore cosmic and global contexts. However, this transition poses challenges for those who staunchly defend "traditional values" and attempt, with diminishing success,

---

[28] Without looking up from the text of the book, I want to draw the reader's attention to the incredibly important context of the trend of the future: we are talking about a purely intellectual component; the sensual component of the Pisces era will erode for the next 500 years in the most active way.

to impose their behavioral norms from the Pisces era[29]. From abortion bans to exclusive club rules in business, these attempts will face growing resistance.

During the transition period, there will be a coexistence of elements from both eras. Online sermons and virtual clubs are examples of such hybrids. While a peaceful transition between eras is desirable, certain paradigms will prove irreconcilable. Contentious issues such as the right to abortion, which traverses the dividing line between the two eras, and totalitarianism, with its vertical system suppressing freedom of expression and individuality through intellectual means, will continue to be points of conflict and debate. It is crucial to recognize that the phrase "so historically happened" will hold little weight in the coming era, as common sense prevails and demands a reevaluation of outdated norms and values.

The storming of the Bastille[30] in Paris stands as a pioneering avantgarde moment of the new era. Through the destruction of this potent symbol of the king's omnipotence, the French demonstrated their perceptive awareness of the changing tides in the world[31]. Renowned for their practicality, they chose this act to express their disdain for symbols and the impotence of the old world.

---

[29] Tesla has the following mission: to create a world that operates with the help of electrical energy, powered by batteries and transported by electric cars.

[30] *Taking Bastilia* (fr. *Prise de la Bastille*) is one of the central episodes of the French Revolution, the storming of the fortress of the Bastille on July 14, 1789. The fortress was built in 1382. It was supposed to serve as a fortification on the outskirts of the capital. Soon it began to serve as a prison, mainly for political prisoners. For 400 years, there were many famous personalities among the prisoners of the Bastille. For many generations of the French, the fortress was a symbol of the omnipotence of kings.

[31] I invite the reader to highlight this point and, after reading the entire book, think about why the French respond so well to the subtle vibrations of change in the world.

Another illustrative example of such impotence can be found in the declining Russian empire, often referred to as the Great Horde or mistakenly labeled as the Golden Horde. Despite its final demise in the 16th century, it intermittently resurfaced as a decaying golem, persisting for several decades before succumbing to decay once again. This cyclic revival of the golem represents a doomed existence[32], gradually losing its vitality compared to the dynamic evolution of the wider world. This phenomenon echoes the contextual changes on a planetary scale.

The Age of Pisces, representing the Old World, is characterized by various counterbalances to its inherent qualities of compassion. One prominent aspect is the prevalence of addiction, particularly to substances such as alcohol and drugs, which serve as means to escape or avoid reality[33]. It is noteworthy that even the symbol of the fight against alcohol addiction, the Society of Alcoholics Anonymous[34], carries the main message of the Age of Aquarius: the rejection of individualism in favor of the collective well-being of society.

Another destructive aspect of Pisces is manifested in destructive materialism, which takes the form of masochism, guilt, propaganda, dogmatism, intolerance, and blind faith. These manifestations have permeated various realms, including science, where cosmological theories

---

[32] "Another day... And one more night.
And it seems that I can't get over it anymore.
Not to endure the weight of non-existence,
And the monotony of the undead."

[33] One example of the transition period is the move of pilgrims — people fleeing rigid religious norms — to America, which, however, contains elements of escapism characteristic of the previous era.

[34] Strictly speaking, the OAA is another hybrid of two bordering eras that uses the divine context of "Let Go and Let God". As they themselves explain, it is the need to admit one's mistakes to a higher power, which corresponds to the context of the Age of Pisces, and the context of anonymity itself is a central aspect in the Age of Aquarius.

and hypotheses have been influenced by such destructive tendencies. Even the intriguing yet still elusive string theory has been subject to a similar negative influence.

In the new world of Aquarius, we will confront the merging of opposites and confront evil directly. There will be no acceptance of evil as a gray area; it will be seen as clearly distinct from what is good. This revelation will challenge the proponents of *realpolitik*[35], as the outdated mindset of relying on uneducated individuals will no longer be fashionable. Young leaders will no longer dare to rely on ignorant fringe groups.

The context of the new element, air, emphasizes rational thinking, which distinguishes itself from realpolitik by its ability to consider alternatives and avoid extreme measures, even the use of excessive force. Realpolitik, on the other hand, is a pseudo-intellectual impotence in the face of cowardly criminals. By surrendering to such individuals, realpolitik only exposes its own foolishness.

During the Age of Aquarius, the primary social conflict revolves around the clash between democracy and fascism. This is not a matter where compromises are appropriate, and those who hold a different perspective should reflect on the Holocaust as the epitome of classical fascism in global politics. It is important to remember this in a broad sense, drawing inspiration from the famous statement by Martin Niemöller[36].

---

[35] *Real policy* (in German. *Realpolitik*; in Russian-language texts, it is often used without translation (*Realpolitik*) or in the form of transliteration *Realpolitik*) — a type of state policy that was introduced and implemented by Bismarck and was named by analogy with the concept proposed by Ludwig von Rohau (1853). The essence of such a course is the rejection of the use of any ideology as the basis of the state course. Such a policy is based primarily on practical considerations, not ideological or moral ones.

[36] "*When they came...*" is a quote from the speeches of the German pastor Martin Niemöller, with which he tried to explain the inaction of German intellectuals and their non-resistance to the Nazis.

THE FOUNDATIONS OF A NEW BUSINESS ETHIC

According to Sandy Penny, the sign of Aquarius encompasses five essential elements: communication, creativity, cooperation, compassion, and society[37]. In contrast, the sign of Pisces is associated with hierarchy, structure, power through control and submission, as well as deep emotions and various forms of addiction and emotional dependencies in relationships[38]. However, it is crucial to understand that these abusive ties, whether in family or social contexts, will be dismantled by the natural shift in the planet's gravity brought about by the precession. It is futile to argue against this cosmic phenomenon, as it will not occur again for approximately 25,000 years. Those who cling to traditionalism may opt to preserve themselves in the hope of resurrection after 12 precessional cycles.

The Age of Aquarius heralds a transition from hierarchical and blood ties to meritocracy and partnerships based on fair assessment of individual contributions. The primitive symbol of limitless communication, the Internet, will be surpassed by more advanced forms of human interaction. The possibilities are vast, ranging from the development of telepathic communication to innovative methods of exchanging information in business, such as replacing traditional financial intermediaries like banks with systems that eliminate unnecessary, meaningless, and costly intermediaries.

---

[37] Sandy Penny's famous original quote is: *"the sign of Aquarius includes what I call the 5 Cs: Communication, Cooperation, Creativity, Compassion and Community"*.

[38] *Karpman Triangle* (Eng. *Karpman drama triangle*), also *Dramati Treugo* (*Drama Triangle*), *Treugólnik Destiny* is a psychological and social model of human interaction in transactional analysis, first described by Stephen Karpman in 1968 in his article Fairy Tales and Script Drama Analysis. This model is used to treat patients in psychology and psychotherapy.

Recognizing that the dismantling of hierarchical relationships and outdated structures primarily targets vertical systems that no longer align with new realities and fail to generate proportional surplus value is crucial for individuals entering the new era. This understanding can help them avoid unnecessary resource loss. However, despite the positive impact of these new realities of free partnership on humanity, the process can be challenging and met with resistance from decaying entities. In such circumstances, honesty becomes the antidote to the imminent danger, but I will delve into this topic in detail later in the book.

In the broader context, the global transition to the Age of Aquarius gives rise to immediate trends, which can be categorized into three main groups, albeit with some degree of conventionality. It is worth noting that progressive development unfolds in the following sequence:

Robotization: The increasing integration of automation and artificial intelligence into various industries and sectors, transforming the nature of work and productivity.
Innovation: A focus on continuous innovation and advancements in technology, science, and creativity, driving societal progress and pushing the boundaries of human knowledge and capabilities.
Humanitarian Ideals: A growing emphasis on humanitarian values and principles, including compassion, social justice, and sustainable development, to address pressing global challenges and create a more equitable and harmonious world.

These trends represent the evolving landscape of the Age of Aquarius and provide a glimpse into the transformative forces shaping our future.

The conventionality of these trends stems from the fact that they may not align with the current state of consciousness and can be challenging to fully comprehend. For instance, the concept of unconditional income may not be universally embraced by society, and concerns such as the bread of shame[39] remain valid. However, these trends set the priorities for attention and resource allocation as per the priorities shown in the pyramid of Maslow concept[40], reshaping the societal architecture around them, despite their imperfections in relation to an abstract standard of development.

Robotization signifies a shift in the social role of individuals, moving from being mere production units to becoming designers and teachers. These two will play a central role in the near future[41]. It raises questions about the balance between collective endeavors and individuality in society.

---

[39] *The bread of shame* (Heb. בושה של לחם) — here in the narrow sense — the shame of receiving a free gift to those who do not deserve it. In this book, much attention will be paid to this concept, but in this narrow sense, we are talking about the concept of calling humanity to eliminate the Bread of Shame in an effort to become a Creator "in the image of God."

[40] *Pyramid of Needs* is the common name for the hierarchical model of human needs, which is a simplified presentation of the ideas of the American psychologist Abraham Maslow. Steps (from bottom to top):
1. Physiological needs;
2. The need for security;
3. The need for love / belonging to something;
4. The need for respect;
5. The need for knowledge;
6. Aesthetic needs;
7. The need for self-actualization.
Moreover, the last three levels: "cognition", "aesthetic" and "self-actualization" are generally called the "Need for self-expression" (Need for personal growth).

[41] There will also be various types of crafts as integrated production, as well as entrepreneurship, and the most valuable will be art and science.

Innovation is synonymous with progress and the era of scientific breakthroughs. It encompasses the transition towards renewable energy sources, the diminishing influence of religious dogmas, the decline of fascist societies, and the dismantling of contexts based on abuse of power.

Humanitarian ideals promote the coexistence and brotherhood of peoples. They involve rejecting the influence of secret societies and sacred contexts, advocating for transparency, and facilitating the public exchange of information.

As the world navigates these transformative trends, society will need to grapple with the challenges and opportunities they present, redefining the foundations of our social structures and striving for a more inclusive and enlightened future.

In the bustling city of New York, where parking is notoriously challenging, over 7,000 places of worship stand tall — churches, synagogues, mosques, and other sacred establishments — each with their respective clergy who hold certain preferences regarding this delicate matter. One can only hope that these religious leaders will embody the principles they preach and relinquish their unwarranted and excessive privileges. Alas, history has shown that such expectations are not always met. Our only recourse is to place our trust in the power of public pressure, for its substantial presence will serve as a tangible measure signifying the onset of an active phase of transformation in societal consciousness during this new era. This transformation aims to redefine the role of religions in human life, urging them to align with the evolving ethos of the times[42].

---

[42] Since this question has probably already arisen in the reader's mind, I will say right away: I am not an atheist for the simple reason that not only I once had a chance to become aware of him/her/it, I personally

THE VERTICAL CONTROL SYSTEM OF THE AGE OF PISCES VERSUS THE HORIZONTAL SYSTEM OF THE AGE OF AQUARIUS

In the present era, the Internet stands as a true nemesis to modern dictators, despite its imperfections. It exemplifies a remarkable network of horizontal connections and undoubtedly belongs to the Age of Aquarius. The antiquated remnants of vertical hierarchies, symbols of a dying world, are being gradually replaced. This transformative process gains momentum through the advancement of communication technologies that defy restrictions. The Age of Pisces, characterized by its emphasis on money, power, hierarchical structures, and stringent control over individual choices, is fading away into obscurity. The Age of Aquarius, in contrast, directs attention towards matters of equality and collaboration.

For instance, the significance of leadership in business ventures is diminishing, giving way to the importance of teamwork. In this context, the exposure of pseudo-leadership around the globe manifests in a grotesque manner. Former dictators have transformed into laughable jesters, while individuals who once weaved carpets have emerged as genuine leaders, subduing their egos for the betterment of a grateful humanity. Moreover, amidst the crumbling systems that safeguard secrets at all levels, the dawn of blessed times marked by true equality among humankind draws near.

Technology in the Age of Aquarius

In the Age of Aquarius, the advancement of technology serves as a complementary force to the societal movement

---

know (and this is exactly the opposition of a belief, in other words of a religion preached by atheists) that God is in everything around us, us inclusive. Therefore, absolutely no intermediaries are needed for dialogue with G d. This is despite the fact that the aspect of dialogue with G d is not at all obvious to me in the context of necessity.

towards information discovery, exchange, and the overall democratization of global order. Disregarding the correlation between technological progress and the liberalization of society would, at the very least, result in the subjugation of the advanced and progressive segment of society by the aggressive and regressive elements. Consequently, such a scenario inevitably leads to the deterioration of scientific advancements and cyclical degradation of society. An illustrative case is China, which, after experiencing over 500 years of historical development under conditions of totalitarianism, has transformed from a politically, economically, and socially advanced nation into a stagnant region unable to prevent external control without the occurrence of a nearly spontaneous civil uprising[43].

Undoubtedly, blockchain technologies, exemplifying initiatives emerging from the bottom of the control hierarchy, and their private application, cryptocurrencies, will initially compete with traditional investment assets. However, this competition is only temporary and stems from the profound imperfections and imbalances within the banking industry. These flaws are not merely historical accidents but deliberately crafted and tenaciously supported by the complacent opposition — short-sighted, cunning, and yet deterministic elders across generations. These individuals possess a unique power that drives them to do nothing while reaping all the world's benefits, unnecessarily and obligatorily.

In the future, banks will find their place in the annals of industries that once marked slaves with their owners'

---

[43] *The Opium Wars* — military conflicts in China in the XIX century between the Western powers and the Qing Empire. One of the main reasons for the hostilities was disagreement over the import, first of all, of raw opium and opium to China, which was literally experiencing an epidemic of drug addiction. Hence the wars got their name.

brands and repaired broken tableware. Blockchain technology, representing the epitome of a democratic state, will undoubtedly revolutionize the financial sector, transcending geographical boundaries and empowering all individuals equally. This technology will liberate the concept of value and means of circulation from financial conglomerates that have accumulated their wealth through extracting value from the circulation of money. These conglomerates are vertically integrated and intricately entwined with the existing state architecture.

As a result, resources generated by humanity will be liberated[44] to invest in venture capital technologies, further accelerating the trend towards innovation. The financial sphere will experience the emancipation of value equivalence and means of circulation, fostering a new era where the possibilities of peace and prosperity are accessible to all.

The individuals I mentioned earlier, possessing a wealth of experience in successful behavioral strategies that have resulted in colossal social imbalances, have effectively transmitted these imbalances through generations, particularly to developing countries along the food chain. Simultaneously, they have instilled in the minds of similar idle individuals, often their own parents, a fervent desire for economic security at any cost, completely disregarding personal development and growth. To justify these disparities, theories of racial, national, and social superiority are diligently fabricated, eagerly embraced by the leading business schools worldwide. These institutions, instead of teaching ethical business practices, set trends in price manipulation and competition restriction.

---

[44] Excessive accumulation (savings) of resources is itself a form of fear for the future, or, in other words, fear that the world is imperfect. But the world is perfect!

As communication technologies advance and travel becomes more accessible, manipulating groups of people becomes increasingly intricate. The traditional chains of control, which have perpetuated centuries of unjust segregation based on gender, race, social status, and other differences, along with practices like banking and insurance scoring, food production licensing, transportation organization, and various forms of fascism, are now subject to unprecedented pressure from new technologies. Moreover, these chains are being challenged by the principles people aspire to live by.

It is crucial to acknowledge that the demand for specific technologies arises from people themselves. The pleas for technology accessibility stem from the suffering of individuals in countries with fascist ideologies or those threatened by dictatorship's grasp. Thus, the lamentations of various Fuhrers about the open architecture of the internet for global communication appear comically absurd. However, dictators never desire to suffer alone; they seek to ensnare the very people they have oppressed. It becomes evident that they are merely another category of losers who have failed to find their own path in life, opting instead to appropriate other people's dreams as their own life plan.

> The concept of work and its nature have undergone a significant transformation. Permanent employment is no longer considered a universal rule or a necessity.

Modern society is increasingly accepting and supportive of individuals' aspirations to develop their talents and pursue self-realization. The primary motive for choosing an occupation is no longer solely focused on survival. Instead, people are driven by a desire for personal fulfillment and meaningful engagement in their work. There is a general trend of diminishing dependence on large corporations, as individuals seek greater independence and autonomy.

The rise of entrepreneurship and venture creation will become increasingly prevalent across the globe. More people will embark on their own entrepreneurial journeys, taking charge of their professional lives and pursuing their passions. As a result, the disparity in wages based on geographic location will continue to diminish, fostering a more equitable distribution of income worldwide.

> Social media serves as a prominent indication of the progressing Aquarius society.

Social media has emerged as a notable manifestation of the evolving Aquarian society. The principles of Aquarius, such as liberation, freedom, and the expression of intellect, align with the goals of social media platforms. These platforms facilitate the development and sharing of knowledge, uncovering the talents of individuals across various strata, races, nationalities, genders, and more. They embody the Aquarian ideal of supporting a diverse and inclusive community.

In the Age of Aquarius, the central value of respecting each person's uniqueness, talents, capabilities, and intelligence holds great significance. Social networks, as a result, are able to undertake entirely new roles that can benefit humanity as a whole. They enable the amplification of public interest and shift away from serving narrow groups' interests. Systems that were created to cater exclusively to a select few in society, rather than benefiting the entirety of society, will no longer be in line with the prevailing trends of the Aquarian age.

> There is a notable decrease in the emphasis placed on power and wealth within society.

There is a noticeable shift away from the excessive emphasis on power and wealth, and a growing emphasis on the distribution of valuable knowledge that benefits all of humanity. The trend is leaning towards recognizing the inherent right to life and the development of individual talents within diverse and self-organized societies. The key factor in fostering tolerance towards such societies lies in acknowledging and appreciating their unique capabilities and contributions. This change reflects a broader societal recognition of the importance of equitable opportunities for personal growth and the value of collective well-being over the pursuit of power and monetary accumulation.

The transition from the age of Pisces to the age of Aquarius is expected to bring about significant upheavals and transformations across various aspects of society.

The transition from the age of Pisces to the age of Aquarius is marked by significant shifts in resource allocation, power dynamics, and economic interests. This transition challenges the established world order of the previous era, leading to its gradual weakening. In line with the second law of thermodynamics, closed systems tend to experience an increase in entropy, implying that various systems will undergo periods of chaos and eventual collapse.

The Age of Aquarius represents a profound exploration of our true selves and a heightened awareness of our spiritual needs. This exploration necessitates a reassessment of values, principles, and ways of life. To gain insights into principles that will be crucial for mankind in the coming 2,000 years, we can turn to the experiences of medieval Japan.

Drawing from that experience, the following principles can guide individuals in surviving and maintaining their sense of self:

1. *Embrace Adaptability*: Cultivate a flexible mindset and the ability to adapt to changing circumstances.
2. *Foster Resilience*: Develop inner strength and resilience to navigate challenges and setbacks.
3. *Emphasize Harmony*: Seek harmony within oneself and with the surrounding environment, promoting balance and peaceful coexistence.
4. *Cultivate Self-Discipline*: Practice self-discipline to maintain focus and achieve personal growth.
5. *Honor Tradition*: Respect and learn from the wisdom of past generations while embracing new ideas and innovations.
6. *Cultivate Mindfulness*: Foster present-moment awareness to enhance clarity and make conscious decisions.
7. *Embrace Simplicity*: Find contentment in simplicity and avoid excessive materialism or attachment.
8. *Nurture Connection*: Prioritize meaningful relationships and connections with others, fostering a sense of belonging and support.

By integrating these principles into our lives, we can navigate the upcoming upheavals and stay true to ourselves amidst the changing times.

Keep yourself engaged and prepared, as this book has the potential to significantly transform your life, just as it did for me. Stay open to the insights and wisdom it offers, as it can be a catalyst for profound personal growth and change.

Chapter II

Samurai and Ronin

The samurai and ronin are both integral parts of Japanese history and feudal society.

Samurai refers to the noble warrior class in medieval Japan, known for their exceptional combat skills and strict adherence to the code of bushido[45]. They served as retainers to feudal lords, known as daimyo, and were expected to display unwavering loyalty and obedience to their masters. The samurai lived by a hierarchical structure

---

[45] *Bushidó* (Jap. 武士道 Bushi-do:, "The Way of the Warrior") is a code of a samurai, a set of rules, recommendations and norms of behavior of a true warrior in society, in battle and alone with himself, a military philosophy and morality rooted in ancient times. Bushido, which originally arose in the form of the principles of a warrior in general, thanks to the ethical values and respect for the arts included in it in the XII–XIII centuries, with the development of the samurai class as noble warriors, grew together with it and finally formed in the XVI–XVII centuries as a code of samurai ethics. The word "bushido" consists of three hieroglyphs. The first two make up the word "bushi" — the only word of several available in the Japanese language for the concept that most accurately conveys the essence of a warrior.

In the first hieroglyph "*bu*", with the meaning "military" / "military", its key is a hieroglyph with the meaning of "stop". And the second part of the sign is an abbreviated version of the ideogram denoting "spear". The ancient Chinese dictionary of Shu Wen gives the following explanation: "Bu is the ability to subdue the weapon and, therefore, stop the spear." In another ancient Chinese source (the Book of Zi Chuan) we find a more detailed interpretation, which says, that bu includes boone, that is, literature, calligraphy, and in a broader sense all non-military arts. Boone prohibits violence and subjugates weapons — "stops the spear".

The hieroglyph "*shi*" in modern Japanese means "military", "warrior", "man" and even "noble man". And initially, in China, this word was defined by people who had skill in a certain field and took their position in society thanks to scholarship, but were ready to take up arms when necessary. Thus, bushi is a person who is able to preserve peace both with the help of art, and by military means.

The third hieroglyph "*do*" — denotes the Path — the most important concept for most Eastern philosophical teachings, in this case uniting these seemingly incompatible qualities — bu and boone, in the way of life of the "ideal person".

and followed a strict code of honor, exposing virtues such as bravery, loyalty, and self-discipline. Their purpose was to protect their lord, maintain social order, and uphold justice.

On the other hand, ronin were samurai who found themselves without a master or lord to serve. This could occur due to various circumstances such as the death or downfall of their lord or their own disillusionment. Ronin were often skilled warriors seeking new employment, but their masterless status left them in a state of social and economic uncertainty. Some ronin would engage in freelance work or become mercenaries, while others pursued different paths, such as becoming scholars or artists.

The ronin were seen as figures of both admiration and caution in Japanese society. They were known for their independence and often portrayed as individuals who took personal responsibility for their actions. While some ronin were revered for their adherence to the samurai code and continued to embody the ideals of bushido, others became outcasts or engaged in unlawful activities.

The contrast between the samurai and ronin reflects the complexity of feudal Japan, with its social structures, honor codes, and the individual journeys of those within it. Both the samurai and ronin played significant roles in shaping Japanese history and continue to be subjects of fascination and study.

Medieval traditions depict the samurai as unwavering warriors, devoted to the principles of bushido, which is often inaccurately referred to as the samurai's code of honor. However, this description is not merely metaphorical. In reality, the samurai lacks a personal code or true honor. Their entire existence is dedicated to their master, who holds authority over their well-being and even life and death. It is essentially a form of slavery deeply rooted in a culture of unwavering hierarchical obedience, characteristic of the Pisces era.

Ironically, the original code of honor for the samurai was written by a ronin[46], individuals who were once samurai but became masterless. For centuries, the samurai revered and adhered to this code as if it were a sacred doctrine, almost akin to religious dogma. This is not surprising since true creators are often found among those who are internally free. It is worth noting that the samurai's moral code[47] developed in parallel with the shogunate system, a harsh form of dictatorship. Like

---

[46] *Yamamoto Tsunetomo* (Jap. 山本 常朝 June 11, 1659 – November 30, 1719) was a samurai from Saga, Hizen, where daimyō Nabeshima Mitsushige ruled. Yamamoto devoted 30 years to serving the daimyo and clan. When Nabeshima died in 1700, Yamamoto did not commit suicide as a sign of loyalty because his ruler had opposed the practice during his lifetime. After disagreements arose with Nabeshima's successor, Yamamoto went into seclusion in the mountains. Later (between 1709 and 1716), he recounted many of his thoughts to his samurai friend, Tsuramoto Tashiro. Many of Yamamoto's aphorisms concerned his father and grandfather (Nabeshima Naoshige), his daimyō. Yamamoto's sayings were collected and published in 1716 under the title Hagakure.

[47] Samurai morality was formed in general terms at the same time as the shogunate system, but its foundations existed long before that time. Nitobe Inazo singled out Buddhism and Shintoism as the main sources of bushido, as well as the teachings of Confucius and Mencius. Indeed, Buddhism and Confucianism, which came to Japan from China along with its culture, had great success with the aristocracy and quickly spread among the samurai. What the samurai lacked in the canons of Buddhism and Confucianism, was abundant in Shintoism.

The most important principles of bushido were drawn from Shintoism — the ancient religion of the Japanese, which was a combination of the cult of nature, ancestors, faith in magic, the existence of the soul and spirits in the things and objects surrounding man, love for the country and the sovereign. Borrowings from Shintoism, which were adopted by bushido, were combined into two concepts: patriotism and loyalty.

other censored societies[48], the seeds of freedom took a long time to sprout, with changes in morality gradually evolving within repressed closed societies. These moral transformations were often presented as religious dogmas, facilitating their dissemination within closed societies.

Behind this intricate moral framework, initially as a mere shadow and eventually as a myth and a deliberate act of mighty warriors, stood the ronin — free individuals who took full responsibility for their actions and choices.

In medieval Japan, the moral framework of warriors, who possessed both martial and artistic prowess, was shaped through revered literary works[49]. These works served as codes of conduct, offering examples of life situations, rhetoric, and decisions made by great warriors. The bushi, or warriors, embodied a unique fusion of

---

[48] Bushido was particularly strongly influenced by Mahayana Buddhism, which penetrated Japan in 522. Many of the philosophical truths of Buddhism most fully met the needs and interests of the samurai. This is, first of all, a reverent attitude to death and indifference to life, based on the belief in the rebirth of souls. At the same time, the most popular sect of Buddhism was "Zen", whose monks made a significant contribution to the development of bushido.

[49] During the reign of Tokugawa Ieyasu, the "Code of Samurai Births" ("Buke sho hatto") was compiled, which determined the norms of behavior of the samurai in the service and in his personal life. The second work devoted to the chanting of the tenets of bushido was a hagiographic description of the exploits of the daimyo Takeda Shingen in twenty volumes, the authorship of which was shared by Kosaka Danjo Nobumasa and Obata Kagenori. Somewhat later, the work of Daidoji Yuzan (1639–1730) "The Initial Foundations of Martial Arts" ("Budo Shoshin Shu"). And finally, in 1716, 11 volumes of the book "Hidden in the Foliage" ("Hagakure") were published, which became the "holy scripture" of bushi. This curious work belonged to Yamamoto Tsunetomo, a former samurai of the Saga clan from the southern island of Kyushu. After the death of his master, daimyō Nabeshima Naoshige, whom he faithfully served for ten years, Yamamoto became a monk and devoted the rest of his life to summarizing the tenets of samurai honor.

art connoisseurs and defenders against evil, willing to confront and halt evil themselves, as symbolized by the concept of "stopping the spear." In essence, this represents a timeless human aspiration: the desire to possess the divine ability to choose between good and evil, and to actively choose to do good just like God[50].

Paolo Coelho, through his exploration of one of his books[51], delves into captivating and thought-provoking arguments about the nature of the warrior of light. Although not presented in a systematic manner, these insights offer intriguing perspectives. However, the reasons why everyone looks up to the warrior of light remain unmentioned.

The very concept of the warrior of light evokes a dreamlike image of an individual who exists between two epochs, characterized by emotional depth and an enigmatic presence. This image carries an inherent

---

[50] Parashat Bereshit, Genesis 1:1–6:8 — the very first chapter of the Torah tells about this vow of a person to G-d.

[51] "The Book of the Warrior of Light" (port. Manual do Guerreiro da Luz) is a book by Brazilian writer Paulo Coelho, published in 1997. The main highlighted thoughts in his book:
A Warrior of Light is a person worth looking up to. But it is only through long and persistent efforts that you will attain such spiritual power
Everything in the world is harmonious. It is interconnected and without one thing there will be no other. (A theory called chair — one leg is unstable and the whole chair will fall)
When you achieve harmony with nature, harmony with people automatically overtakes you
Experience is the only pillar of life
There are no ideal and imperfect people. There are not completely righteous people
Do not react to external factors. What's inside is much more important (that is, if you are called names, do not react to it hot-temperedly, it's just a waste of energy, a smart person will always prove it in practice)
Cm. Thought 2. Friendship is the complementarity of each other with certain qualities, that is, "together friends are a warrior of light" ...

longing for respect, reflecting a remnant of the age of Pisces while also encompassing elements of the new era. It can be viewed as an intermediate variation, arising from its detachment from the rigid confines of everyday life.

This image of the warrior of light holds a certain allure due to its ability to transcend the absolute and mundane aspects of existence. It captivates us by representing a bridge between different contexts and offering a glimpse of something beyond the ordinary.

In our case, it appears that the method available to us is for the warrior of light to prevent the spread of evil beyond the harm already inflicted upon them. Through the exploration of their books and engaging in intriguing discussions about the nature of the warrior of light, we come to realize the reasons why individuals aspire to emulate them. This dream-like image represents a person standing between two epochs, eliciting an emotional and unproven desire for respect — a reflection of the characteristics of the age of Pisces, while also encompassing elements of the new era.

Again, I perceive this image as an intermediary concept, partly detached from the absolute and mundane context of everyday life. I firmly believe that spirituality serves to fulfill our task of personal growth in the physical world, here and now. This growth does not require enlightenment as a prerequisite; rather, our battles and lessons are learned directly from our daily lives.

Over the course of several centuries, the perception of the ronin has undergone significant changes. Initially depicted as bandits and rapists, the image transformed into that of a warrior who willingly abandoned[52] a relatively

---

[52] There was even a textbook saying "Seven Falls, Eight Rises", which meant the right of a samurai to wander for a period of one year seven times during his service, each time returning to the service of his patron. In the context of samurai service, this meant that the master could be patient and take his slave back, but there was also a limit

secure but dependent existence in exchange for a short life marked by conscious freedom. During the 250-year Edo period[53], under the strict class system of the shogunate, the number of ronin increased substantially.

This increase can be attributed to the societal shift in Japan from dictatorship to totalitarianism[54]. Totalitarianism, characterized by the extreme suppression of freedoms, became the catalyst for the destruction of the old system and the dawn of a new stage in societal development. This era witnessed a flourishing of art, literature, and the emergence of public opinion as a social institution. Restrictions were placed on samurai, prohibiting them from seeking employment with other samurai, and interclass marriages were banned.

Being a ronin during this time was highly undesirable, as it often meant resorting to robbery or relying on charity for survival. The temptation to use force to sustain oneself was great, particularly for warriors who had been raised to embrace a disregard for death. It was only after reaching a spiritual level of selflessness that tipped the scales towards acts of mercy[55] that a ronin could overcome the allure of violence for sustenance. Consequently, within the vertically integrated samurai society, the figure of the ronin was portrayed as a loser, stripped of honor due to circumstances or personal choices. The sole honor and

---

to the master's "peacefulness" is a cynical label of a slave-owning society.

[53] *Period Édo* is the historical period of Japan, the reign of the Tokugawa clan. It began with the appointment of Tokugawa Ieyasu as shogun in 1603. It ended in 1868 with the resignation of the shogun Tokugawa Yoshinobu.

[54] *Totalitarianism* (fr. *totalitarism*) A type of political regime characterized by complete state control over all spheres of society, the elimination of rights and freedoms, repression of the opposition and dissidents.

[55] Like a warrior who can simply rob anyone under the threat of force.

dignity of a samurai lay in their unwavering service to their master.

Behind the mockery of samurai towards ronin lay a latent envy of their courage to choose freedom over a life of material prosperity and imagined success. Legends abound about simple ronin who protected impoverished villagers from arrogant samurai, who would kill for a perceived lack of respect or even without reason, engaging in the demonic practice of tsujigiri[56] — randomly targeting and slaying passers-by at night. Kabuki theater, in particular, has celebrated the story of the 47 ronin[57] who fulfilled their destiny at the cost of their own lives, as they understood it.

---

[56] *Tsujigiri* (辻斬り or 辻斬, literally: murder at the crossroads, eng. *Tsujigiri*) is a Japanese practice in which a samurai, having received a new katana or developing a new type of combat or weapon, tested its effectiveness by attacking a random opponent, usually a random defenseless passerby, in many cases at night. Practitioners of this were also called tsujigiri.

[57] "*Ako's Revenge*", "*Forty-seven ronins*" (Jap. 赤穂浪士 Letters. "Wandering Samurai of Ako"); Forty-seven Samurai) is a Japanese folk legend based on real events, telling about the revenge of forty-seven former samurai for the death of their master.
The story tells how forty-seven ronin prepared and carried out a plan to avenge Kira Kozuke-no-Suke, the court of the Tokugawa Shogun Tsunayoshi, for the death of their master, the daimyo Asano Takumi-no-Kami Naganori of Ako. In 1701, Asano was sentenced to seppuku for attacking a courtier in response to insults and bullying by the latter.
After losing their master, forty-seven ronin, led by chief advisor Oishi Kuranosuke (Japanese. 大石 良雄 Ōishi Yoshiō, title 内蔵助, Kuranosuke), swore an oath to avenge death for death, despite the fact that they were sentenced to death for it.
In order not to arouse suspicion, the conspirators disappeared into the crowd, becoming merchants and monks, while Oishi moved to Kyoto and began to lead a riotous lifestyle, divorced his wife and took a young concubine. Eventually, after learning that the ronin had scattered and Oishi was drunk, Kira loosened his guard and became more careless.
Meanwhile, the ronin secretly collected and transported weapons to Edo, gaining the trust of Kira's household (one of Asano's former

servants even married the daughter of the builder of the official's estate to get plans for the construction).

When everything was ready for the execution of his plan, Oishi secretly moved to Edo, where all the conspirators met and reswore an oath of revenge.

In the 15th year of the Genroku era, on the 14th day of the 12th month (January 30, 1703), the ronin attacked Kira's estate in Edo with two squads at a drum signal, killing 16 and wounding more than 20 people. Kira managed to hide in the house with the women and children in a large closet, and they could not find him for a long time. However, Oishi, after checking Kira's bed, made sure that it was still warm. Soon, behind the wall scroll, a secret passage was discovered, leading to a hidden patio with a small storage structure for storing coal, which was protected by two armed guards. Kira was found there. Oishi respectfully told him that they, Asano's former servants, had come to avenge their master. As a samurai, Kira was respectfully asked to commit ritual suicide, but he refused or simply could not do it. Then Oishi himself killed Kira by cutting off his head.

The head of the defeated enemy of the ronin was carried to the monastery of Sengakuji to the grave of his master, thereby fulfilling the oath.

The authorities were in a quandary. On the one hand, the ronin acted according to the letter and spirit of bushido, avenging their overlord; on the other hand, they disobeyed the shogun's order, infiltrated Edo with weapons, and attacked the courtier. Due to the growing popularity of the forty-seven ronin among the people, the shogun received many petitions for them, but, as expected, sentenced all conspirators to death in accordance with the law. However, they were allowed to perform the noble rite of ritual suicide, as befits real samurai, instead of being executed as ordinary criminals.

Seppuku took place in the 16th year of the Genroku era on the 4th day of the 2nd month (March 20, 1703). The lowest rank of the participants in the revenge, Oishi, immediately after its accomplishment, sent home to Ako as a messenger. The forty-six ronin who remained in Edo were buried in the same monastery as their master. Their graves have since become an object of worship, and their clothes and weapons, according to legend, are still kept by the monks of Sengakuji. The good name of the Asano family was restored, and some of the former possessions were even returned to his family. The last of this group of ronin returned to Edo, was pardoned by the shogun, lived 78 years, and was buried next to his comrades.

exemplified the principles of the ronin way of life:

Miyamoto Musashi[58]

Musashi was a legendary swordsman who lived in 17th century Japan. Renowned as one of the greatest swordsmen in Japanese history, Musashi engaged in duels with other samurai. He was also a prolific writer and artist, dedicating much of his life to honing his skills in these domains. Musashi embraced a simple and austere lifestyle, often traveling alone and relying on his warrior and artistic abilities to sustain himself.

Benjamin Franklin

Franklin was one of the founding fathers of the United States, known for his contributions as an inventor, scientist, and writer. His famous aphorisms, such as "early to bed and early to rise makes a man healthy, wealthy, and wise," are still widely quoted. Franklin was a savvy businessman and investor, using his wealth to support causes aligned with his beliefs, such as public libraries and universities.

---

[58] *Miyamoto Musashi* (Jap. 宮本 武蔵 (1584 – June 13, 1645), also known as Shinmen Takezo, Miyamoto Bennosuke, Shimen Musashi-no-Kami Fujiwa-ra-no-Genshin, or by his Buddhist name Niten Doraku, is a Japanese ronin, considered one of the most famous swordsmen in Japanese history. Contemporaries gave him the nickname Kensei (Jap. 剣聖, "Holy Sword"). Musashi became famous for his outstanding sword technique, which he honed from early childhood in many fights using a wooden sword. He is the founder of the school of Hyoho Niten Ichiryu or the samurai art of fighting with two swords Nitoryu (Japanese: Nyōryū). 二刀流). Earth and Sky School. He introduced the concept of bokken as a very real military weapon, and not a training one. He successfully used the technique of fighting with two swords, a long one — a katana and a short one — wakizashi. He wrote "The Book of Five Rings" about tactics, strategy and philosophy of military craft, which enjoys a certain popularity at the present time.

Mahatma Gandhi

Gandhi was a political and spiritual leader who played a crucial role in India's struggle for independence from British rule. He advocated for nonviolent resistance as a means to protest against colonial oppression. Gandhi emphasized the importance of simple living and self-sufficiency, promoting practices like growing one's own food, making clothing, and living in harmony with nature.

These individuals demonstrate the principles of the ronin lifestyle in diverse ways. Each focused on self-development, self-protection, and self-sufficiency, utilizing their skills and resources to achieve their goals and create a positive impact on the world around them.

The question arises as to why the world is structured in a way that allows for such chaotic and unjust circumstances to occur throughout history. The answer lies in the understanding that without the turmoil of medieval dictatorships, there would be no true comprehension of the necessity for power accountability and democracy. Miyamoto Musashi, the greatest among the ronin, provides an example. He declined the comfort offered by the Hosokawa clan and spent two years in solitude, contemplating life. Despite gaining fame as an invincible swordsman at the age of twenty-eight, Musashi did not rest on his laurels. Instead, he immersed himself in the study of the Way of the Sword, evolving from a cruel and stubborn individual to a humble and honest one, focusing not only on improving his technique but also on the comprehension of its spiritual essence. Through his deep immersion in the practice and understanding of his craft, he underwent a profound personal transformation. Musashi's journey allowed him to shed his previous negative qualities and embrace virtues such as modesty and honesty. This evolution demonstrates the potential for personal growth and inner change that can be attained through dedicated self-reflection and the pursuit of a higher spiritual path.

FROM THE PREFACE TO THE BOOK OF THE FIVE RINGS:

Musashi wrote: "If you comprehend the Path of Strategy, there will be no incomprehensible ... You will see the Way in everything." And he himself really became a master of several other arts and crafts. Musashi created paintings that are highly valued in Japan. He painted cormorants, herons, dragons, birds with flowers, birds on a dry branch, Bodhidharma, Hotei and other characters. He was a skilled calligrapher, as evidenced by his scroll "Sanki" ("Spirit of War"). In addition, he was a sculptor and master of engraving on metal. It is said that he also wrote poems and songs, but they have not survived to this day. Legend has it that the shogun Iemitsu[59] himself commissioned him to draw a sunrise over Edojo Castle.

It is evident that the path of humanity involves the comprehension of the fundamental nature of good and evil, leading to a conscious choice to embrace good, with love being its highest expression. This same principle reveals the divine essence of evil, contrasting it with good.

Initially, the status of the ronin was undesirable as they lacked a permanent salary from their masters, which was considered essential for every true samurai's existence. This predicament was both a curse and a blessing for the ronin. The absence of material dependence on a master, coupled with the challenging poverty that often fueled the violent tendencies of skilled warriors, gave rise to a true divine independence among the ronin. They made life choices guided not by the illusory duty of serving a

---

[59] *Tokugawa Iemitsu* (徳川 家光, August 12, 1604 – June 8, 1651) was the third shōgun of the Tokugawa dynasty.

master as a slave but by the aspirations of their souls and the desire to rediscover the divine within themselves.

In Musashi's notable work, "The Book of Five Rings," the essence does not solely lie in the rules of fencing[60] or strategies for large-scale battles. The core focus is on the true spirit and understanding of the Way of the Warrior, which the samurai associated with themselves, even deifying Musashi's image after his death. However, as is often the case, this did not prevent them from harboring animosity towards Musashi during his lifetime.

"The Book of Five Rings" represents a philosophical doctrine and an educational system. For Musashi, nothing surpasses the importance of the freedom to embrace death[61] and possess an unwavering spirit of victory. His belief was that by mastering the art of swordsmanship and overcoming at least one opponent, one could overcome any adversary in the world.

In essence, Musashi's teachings encompassed far more than mere combat techniques. They offered insights into the profound mindset, discipline, and philosophy required for a true warrior, transcending physical battles and encompassing the cultivation of one's inner self and the pursuit of personal excellence.

WHY ARE WE TALKING ABOUT THIS?

We are discussing these principles and characteristics of the ronin because they provide valuable insights into a self-sufficient and fulfilling way of life. Unlike the samurai, who valued blind loyalty and unquestioning devotion,

---

[60] Miamoto Musashi, in my humble opinion, is the greatest of the ronin, as he defeated the most feared of dragons, his own ego: he killed like a scythe of death, believing himself to be the highest of warriors, and then snatched the demon from his chest and became a star of kindness in the firmament over the world.

[61] Pay attention to the context of the granting of freedom in death.

the ronin consciously seeks to buy their freedom from societal constraints. The ronin's awareness is guided by several factors:

Resourcefullness

The ronin understands the importance of prioritizing their own freedom and independence. This entails being self-sufficient and effectively managing their finances and resources. They prioritize their own development through education, skills training, and personal growth. Additionally, they ensure the well-being of their family and loved ones by providing financial and emotional support, as well as creating a supportive environment for them to thrive. Furthermore, the ronin prioritizes charity and supports social initiatives. They disperse their sources of income, aiming for multiple financial streams to avoid dependence on a single source.

Self-development

The ronin recognizes that the growth of awareness is the fundamental reason for existence. They prioritize continuous self-improvement, seeking new knowledge, developing new skills, and expanding their horizons. They remain open to learning from their own experiences and from others. Negative attitudes, limiting beliefs, and bad habits are eliminated to facilitate personal growth. By prioritizing self-development, the ronin enhances their own life and positively influences those around them.

Self-defense

The ronin places importance on physical and mental well-being and ensures their own safety. This involves engaging in sports and exercise, mindfulness practices, and mental health practices. They create a safe living environment and take precautions such as having insurance and regular medical check-ups to protect themselves and their loved ones. By prioritizing self-defense, the ronin ensures a long, healthy, and fulfilling life. Martial arts practitioners understand the delicate balance of being dangerous yet self-controlled. Choosing safety alone can lead to weakness, and the ronin strives to be as dangerous as fire while maintaining a softness like water.

By focusing on resourcefullness, self-development, and self-protection, the ronin can lead a self-sufficient and fulfilling life. These principles enable them to achieve personal goals, ensure their well-being, and have a positive impact on the world around them. As the greatest of the ronin, Miyamoto Musashi, proclaimed, the spirit of victory remains consistent across countless battles, regardless of the type or size of the weapon employed.

Reflect upon the ideas that have been expressed.

Chapter III

Honesty and Madness

I t is crucial to recognize that honesty plays a pivotal role in successful communication. However, it is important to understand that the concept of honesty should be viewed within the context of trade secrets, classified information, and varying levels of access to different layers of knowledge. It is essential to acknowledge that there is no universal "most effective level" of honesty applicable in all situations. The appropriate level of honesty will depend on the specific context and purpose of the communication at hand.

In general, honest communication has a greater likelihood of building trust and yielding positive results, whereas dishonest communication can have detrimental consequences. Nevertheless, it is worth noting that surgical honesty may not always be suitable. For instance, there may be instances when it is not appropriate to share certain aspects and details of the world with a child, as it may disrupt their development and create an unhealthy imbalance. Therefore, it is imperative to carefully consider when honesty is necessary and appropriate in communication. This mindful approach will contribute to effective and constructive exchanges.

The silver lining, or the most advantageous aspect, of honesty lies in its ability to cultivate trust and goodwill in the life of every individual. Honesty also enables the establishment of meaningful and enduring relationships built on mutual trust, respect, and understanding. Additionally, honest behavior can increase productivity and foster a more supportive work environment. Embracing honesty as a guiding principle can bring about positive outcomes and contribute to personal and professional growth.

> Let's highlight the possible positive and possible negative traits of honesty

Positive Traits of Honesty:

1. *Trust and Respect*: Honesty brings trust and respect to your relationships, as it establishes a foundation of transparency and reliability.
2. *Ease and Lightness*: Being straightforward and honest is easier than trying to remember lies, relieving the burden of constantly keeping track of falsehoods.
3. *Moral Character*: Honesty strengthens your moral character, demonstrating integrity and a commitment to ethical values.
4. *Healthy Communication*: Honesty is more likely to foster healthy communication, as it encourages open and genuine dialogue.
5. *Peace of Mind*: Being honest allows you to sleep peacefully at night, knowing that you have acted with integrity in all your actions.
6. *Self-esteem and Happiness*: Honesty contributes to self-esteem and overall happiness, as it aligns with living authentically and being true to oneself.

Negative Traits of Honesty:

1. *Potential Cruelty*: Sometimes, honesty can be harsh and hurtful, especially when delivering difficult truths without considering the impact on others' emotions.
2. *Uncomfortable Situations*: Being too frank can lead to uncomfortable situations, where your honesty may be met with resistance or conflict.
3. *Conflict Potential*: Honesty can occasionally result in conflicts, as differing opinions or perspectives are expressed with candor.
4. *Reputation Risk*: Excessive honesty may damage your reputation, even when unwarranted, as some people may perceive brutal honesty negatively.

5. *Unpopularity of Truth*: Telling the truth when it is unpopular can be challenging, as it may invite criticism or social backlash.

It is essential to find a balance between honesty and sensitivity, considering the potential positive and negative outcomes of our words. By being mindful of both the benefits and drawbacks of honesty, we can navigate our interactions with greater understanding and empathy.

Honesty is highly valued by human morality, and this is reflected in major religions. The Talmud forbids lying or deceiving others, emphasizing the importance of sincerity: *"The Holy One, blessed be He, hates the man who says one thing with his mouth and another with his heart."*[62] It also prohibits fraud in business relationships, emphasizing the significance of keeping one's word: *"Since there is injustice in buying and selling, there are mistakes in words. When a man makes a vow to the Lord or takes an oath to bind himself with a pledge, he must not break his word; he will do whatever comes out of his mouth."*[63]

The Qur'an emphasizes the need for strict observance of justice and being truthful, even if it goes against oneself or close relatives: *"O those who believe! Be strict in observing justice and be witnesses before Allah, even if it is against yourself or against your parents and relatives. Whether he is rich or poor, Allah cares about both of them more than you do. Therefore, do not follow low desires so that you can do what is just. And if you conceal the truth or evade it, then remember that Allah is well aware of what you are doing."*[64]

The Bible encourages rejecting falsehood and speaking the truth to one's neighbor, emphasizing the interconnectedness of humanity: *"Having rejected*

---

[62] Judaism. Passachim 113 b

[63] Judaism, Tanakh, Numbers, 30:2

[64] Islam. Holy Qur'an, 4:136

falsehood, let each one speak the truth to his neighbor, for we are members of one another."[65] Similarly, in Buddhism, the importance of truthfulness is emphasized: "*A person must tell the truth.*"[66]

Even the Indian Mahabharata stresses the significance of truth, suggesting that it is always appropriate to tell the truth and that the truth is of the greatest benefit to all beings[67].

These religious teachings highlight the universal recognition of the value of honesty and truthfulness as essential principles for ethical living and harmonious relationships.

It is important to note that previous research has indicated varying findings regarding the relationship between religiosity and honesty. Some studies suggest that religious individuals tend to consider lying as more morally reprehensible compared to non-religious individuals. This moral objection to dishonesty may influence their behavior, leading to a greater tendency towards honesty.

On the other hand, other research has presented evidence that there is no substantial difference in the level of honesty between religious and non-religious individuals. These studies indicate that factors such as personal values, cultural norms, and individual characteristics may have a more significant impact on honesty than religious beliefs alone.

The complex nature of honesty and its relationship with religious beliefs requires further exploration and investigation. It is essential to consider a diverse range of factors that can influence an individual's honesty, including religious, cultural, social, and personal aspects.

---

[65] Christianity. Bible, Ephesians 4:25

[66] Buddhism, Dhammapada, 224

[67] Hinduism, Mahabharata, Shanti Parva, 329:13

All of today's major religions extol honesty as a vital virtue. The emphasis that religions place on honesty is part of a broader connection between morality and religion — an association that has led many in part to speculate about the morality of religious people. Extensive research has shown that the more religious a person is, the more trustworthy they are. People tend to trust believers of other religions — even religions, which they had never heard of, more than doubting the adherents of their own religion. Decades of research on religiosity and honesty have failed to find any consistent link, except for a robust correlation between religiosity and socially desirable response (SDR). This connection can best be explained by the dispositional tendency to self-deception and positive illusions — the third. a variable that underlies both religiosity and SDR. In fact, this dispositional inclination may have shaped aspects of religion precisely to satisfy the need for such aggrandizement.[68]

One of the key findings of these studies is that deeply religious individuals tend to view lying more negatively than their less religious counterparts. However, the actual links between religiosity and lying are often insignificant and inconsistent. In some cases where a relationship exists, it has been observed that more religious people may be more prone to lying than those who are less religious. Interestingly, individuals who are externally motivated by religious practices, such as social communication and a sense of security, are more likely to engage in dishonest behavior.

Lying carries consequences, as it affects how others perceive and treat the liar. While lying may be addressed in therapy or personal relationships, employers and others may not easily forgive such behavior. Even if one convinces themselves that lying is normal, it still conflicts with the dictates of conscience. Personal experiences have

---

[68] PsycInfo database entry (c) 2020 APA.

shown that even lies with good intentions and meant for salvation can have severe negative impacts, damaging personal relationships.

Drawing a line between brutal honesty and withholding certain details is important. Contextual considerations can guide decision-making, such as assessing potential harm caused by withholding the truth, the potential for positive change through honest feedback, personal feelings if the truth were concealed, and determining whether avoiding the truth is an act of cowardice or compassion. The answers to these questions can provide a contextual algorithm for behavior in each specific case.

> From these insights, several paradigms for applying honesty in life can be derived:

1. *Tell the truth*: Strive to always be truthful and avoid even the slightest lies and exaggerations.
2. *Admit mistakes*: Maintain honesty by acknowledging and apologizing for your mistakes.
3. *Be transparent*: Openly communicate your actions and decisions, allowing others to understand your motivations.
4. *Respect confidentiality*: Honor the trust of others by keeping their confidential information private.
5. *Take responsibility*: Accepting responsibility for your actions demonstrates honesty and integrity.
6. *Avoid cheating*: Refrain from deceiving, manipulating, or misleading others.
7. *Be consistent*: Demonstrating consistency between your words and actions reinforces your honesty and decency.

By embracing these principles, individuals can cultivate a lifestyle of honesty, fostering trust, respect, and positive relationships.

Study this issue carefully.

Chapter IV

Algorithms of Life Choice

1. THE BEST IS THE ENEMY OF THE GOOD

In life, we are faced with choices and decisions that can shape our future. One important principle to remember is that "the best is the enemy of the good." This means that striving for perfection can sometimes hinder progress and lead to missed opportunities.

Imagine you are searching for the perfect partner. You decide to meet with ten potential candidates over the course of a month. But how much time should you spend on each person? According to the "*37% Rule*", you should spend approximately 37% of your time exploring your options without making any commitments. Then, once you have evaluated three candidates, choose the best one so far. From there, continue exploring other options until you find someone better than the first three.

This rule applies not only to finding a partner but also to other areas of life, such as finding a job or renting an apartment. It is a universal principle rooted in mathematical statistics and the theory of optimal stopping. The concept of optimal stopping involves choosing the right time to take action in order to maximize rewards or minimize costs.

Optimal stopping problems can be found in various fields, including statistics, economics, and mathematical finance, particularly in option pricing. The key challenge is striking a balance between exploration and exploitation. Choosing to exploit without proper research carries significant risks, while focusing solely on exploration can lead to lost time and waning interest. To solve optimal stopping problems, *dynamic programming*[69] techniques

---

[69] *Dynamic programming* is basically an optimization of simple recursion. A function is recursive if its definition contains a call to the same function. Recursion is simple if the function call occurs only once in a particular branch. Wherever we see a recursive solution with repeated calls to the same inputs, we can optimize it with dynamic programming. The idea is to simply store the results

are often employed, with the *Bellman equation*[70] being a common tool.

By understanding these algorithms of life choice and the principles behind optimal stopping, we can make more informed decisions and maximize our chances of achieving favorable outcomes in various aspects of life.

The implication of this rule is that the principle of "As above, so below" is a working rule of peace, echoing the Latin phrase *"Quod est superius est sicut quod inferius, et quod inferius est sicut quod est superius."*[71] This phrase, found in the widely circulated medieval Latin translation of the Emerald Tablet, captures the idea that what occurs on a higher level is mirrored on a lower level, and vice versa.

In practical terms, this means that mathematical and physical laws have broader applications beyond their specific domains. The algorithm discussed earlier, based on the 37% Rule and optimal stopping, exemplifies this

---

of the subtasks so that we don't have to recalculate them later when needed. This simple optimization reduces the complexity of time from exponential to polynomial.

[70] *Bellman equation*, named after Richard E. Bellman, is a necessary condition for optimality associated with a mathematical optimization technique known as dynamic programming. It records the "value" of a decision-making problem at a particular point in time in terms of the gain from some initial choices and the "value" of the remaining decision-making problem, which is the result of those initial choices. the optimization problem into a sequence of simpler subproblems, as prescribed by Bellman's "optimality principle". The equation is applicable to algebraic structures with complete ordering; for algebraic structures with partial ordering, the general Bellman equation can be used. In other words, the long-term reward for a given action is equal to the reward for the current action combined with the expected reward for future actions taken at the next point in time.

[71] That which is above is like that which is below, and that which is below is like that which is above.

principle. It demonstrates that statistical principles derived from mathematics can be applied effectively across different areas of life.

By recognizing the interconnectedness of natural laws and their manifestation in various aspects of life, we gain a deeper understanding of the underlying order and patterns that govern our world. Applying these principles to our decision-making processes can lead to more favorable outcomes and a greater sense of harmony between our actions and the broader universe.

In essence, the "so above as below principle" reinforces the idea that the same fundamental principles are at work in different realms. It reminds us to consider the broader implications of natural laws in our daily lives and seek alignment with these principles for greater coherence and success.

2. SMILE INTO THE DARK

Exploring uncharted territories and embracing novel experiences is a pivotal component of individual and vocational maturation. Venturing beyond the confines of familiarity and embracing fresh challenges can induce trepidation, but this audacious pursuit promises thrilling escapades, untapped prospects, and enlightening revelations. This principle resonates particularly within the corporate realm, where adaptability and innovation bear utmost significance for triumph.

Embarking on untried endeavors confers numerous advantages in the business sphere. For instance, experimenting with novel products or services fosters a competitive edge and entices prospective clientele. By embracing fresh initiatives or forging partnerships, one can expand their network and enhance their prominence. Adopting novel problem-solving techniques and decision-making approaches nurtures agility and efficacy in the workplace, catalyzing growth and development. Indeed, it is through such daring expeditions that iconic creations such as champagne and blue cheese were brought to fruition.

Richard Bach's renowned book, A Seagull Named Jonathan Livingston, epitomizes the significance of embracing novelty. The narrative follows the odyssey of Jonathan, a discontented seagull yearning for a life beyond the ordinary. Defying derision and fear from fellow gulls, he dares to transcend his limitations, striving to soar faster and higher than any of his predecessors.

Through relentless determination and unwavering resolve, Jonathan evolves into a consummate aviator, discovering a sublime realm that transcends the physical realm. This fable serves as a potent metaphor, underscoring the importance of pursuing aspirations, irrespective of resistance and apprehension.

While venturing into uncharted territories may evoke trepidation, it remains an integral facet of personal and professional growth. By stepping beyond our comfort zones and embracing uncharted challenges, we expand our competencies, knowledge, and aptitude. The tale of Jonathan Livingston, the audacious seagull, serves as a resounding reminder that daring risks and pursuing dreams can yield profound revelations and experiences that surpass our wildest imaginings.

3. EVEN IN MODERATION, EXCESSIVE INDULGENCE CAN BE DETRIMENTAL[72]

How many times have you felt the urge to meticulously organize everything: notebooks, books on the shelf, papers on the desktop? This inclination often reflects the clutter within one's own mind. It is a recognized truth that

[72] I will deliberately break the rules of citation specifically for this chapter. The expression belongs to Anton Chekhov.

disorder in one's life can lead to widespread destruction, as exemplified by various historical instances.

A notable example of excessive disorder in a personal life, resulting in grave consequences on a global scale, can be found in the case of Adolf Hitler, the leader of Nazi Germany. Hitler's personal life was marked by erratic behavior, including drug abuse, extreme paranoia, and delusions of grandeur. These personal challenges contributed to his irrational decision-making, ultimately leading to the loss of millions of lives during World War II and the abhorrent tragedy of the *Holocaust*[73].

I apologize for any previous frustration caused. If there are any further specific details or phrasing you would like me to incorporate, please let me know, and I'll be glad to assist you.

Hitler's obsession with order and control also played a significant role in the devastating consequences he unleashed on a global scale. He envisioned an ideal and meticulously ordered society based on his distorted ideology, leading to the genocide of millions of Jews, Gypsies, disabled individuals, and others deemed "undesirable" by the Nazi regime.

Another example of excessive order in personal life and its negative impact on a global level can be seen in the story of Joseph Stalin, the Soviet dictator who held power from the 1920s until his death in 1953. Stalin's personal

---

[73] *Holokóst* (from the English. *holocaust*, from ancient Greek. ὁλοκαύστος — "burnt offering") — the persecution and mass extermination by the Nazis of representatives of various ethnic and social groups (prisoners of war, Poles, Jews, Gypsies, homosexual men, Freemasons, hopelessly ill and disabled, etc.) during the existence of Nazi Germany, expressed, for example, in the genocide of the Ukrainian (another shame of humanity is the Holodomor), the Chechen people, the Crimean Tatars, the Chukchi, and with them the Koryaks, Itelmens, Lena Yakuts, Daurs, Tungusians, Yukaghirs, Mantsov, Teleuts, Manchus, Altaians, Kets, Oorochi, Sami, Kereks, etc., etc.

life was characterized by an overwhelming need for control and order. This desire for absolute control resulted in the establishment of a totalitarian state that suppressed individual freedom and stifled creativity.

Stalin's policies of forced collectivization and rapid industrialization caused the deaths of millions due to famine, executions, and labor camps. His unwavering adherence to ideology and intolerance of dissent further hindered innovation and progress in the Soviet Union, ultimately contributing to its decline and eventual collapse.

Both examples demonstrate how personal traits and extreme ideologies can enable leaders to seize substantial power, leading to disastrous consequences for millions of people.

Consider the task of organizing a bookshelf. While arranging books alphabetically or categorizing them by topic or purpose can create a comfortable and orderly space, it's essential to question whether such meticulousness is truly necessary. The time spent organizing should be justified by the future benefits it provides. Avoid wasting time on unnecessary sorting and prioritize efforts that genuinely save time and effort.

4. FORGIVE YOURSELF FOR MISTAKES THAT YOU CANNOT CORRECT AND IMMEDIATELY CORRECT THOSE MISTAKES THAT YOU FIND

Forgiving oneself for irreparable mistakes is a crucial aspect of cultivating self-compassion and fostering personal growth. It is essential to acknowledge that as humans, we are prone to making errors and that these mistakes serve as valuable opportunities for learning. However, it is equally vital to hold ourselves accountable for correctable mistakes and address them promptly and appropriately.

Timely correction of mistakes can prevent others from perceiving them as intentional or malicious, particularly in professional or legal contexts where delays in rectifying errors can lead to more severe consequences.

The Japanese principle of "kaizen" underscores the significance of immediate mistake correction. Kaizen embodies a philosophy of continuous improvement, wherein small, incremental changes are implemented to enhance processes and outcomes. Within this framework, mistakes are regarded as opportunities for growth and improvement, with prompt correction serving as an integral part of the process.

A compelling example highlighting the value of immediate correction can be found in the Japanese art of kintsugi, which involves repairing broken pottery using gold or other precious metals. The philosophy behind kintsugi is that the mended pottery becomes even more exquisite and valuable than its original state, precisely because it has been broken and repaired. Thus, mistakes are transformed into opportunities for beauty and personal development.

To foster self-compassion and personal growth, it is crucial to forgive oneself for mistakes that cannot be rectified. Simultaneously, it is equally vital to take responsibility for correctable mistakes and address them promptly. The Japanese tradition of kaizen emphasizes immediate mistake correction, and the art of kintsugi exemplifies how mistakes can be transformed into opportunities for beauty and growth.

5. PRIORITIZATION OF GOALS

Prioritization of goals plays a pivotal role in successfully accomplishing tasks. When we have a clear understanding of our objectives and the necessary steps to achieve them, we can make progress more efficiently and effectively. An important aspect of goal prioritization involves knowing when to temporarily postpone certain tasks in order to focus on more critical ones.

In the face of difficulties or challenges, it may be tempting to fixate on the problem and invest significant time in finding a solution. However, this can cause us to lose sight of the bigger picture and neglect other essential tasks that require attention. In such situations, it is crucial to prioritize the most important goals related to the task at hand and set aside the stuck problem temporarily.

By taking a break and revisiting the problem later, we often gain fresh perspectives and new ideas that can facilitate a more effective resolution. Additionally, by focusing on other vital tasks, we can make progress towards our goals and build momentum, which can ultimately help us overcome the stuck problem more swiftly and easily.

Project management offers a notable example of the significance of goal prioritization. Project managers frequently utilize tools like the Eisenhower matrix to prioritize tasks and allocate resources efficiently. This involves categorizing tasks based on their urgency and importance, and then prioritizing the most crucial items, even if it means temporarily deferring less significant tasks.

In summary, prioritizing goals is a fundamental aspect of successful task completion. By temporarily postponing stuck problems and giving priority to critical tasks, we can gain momentum, gain fresh perspectives, and overcome obstacles more effectively. The Eisenhower Matrix is one tool that can be utilized to prioritize tasks and allocate resources efficiently. Various approaches, such as the "early deadline" strategy or focusing on the most important tasks, can also contribute to effective planning and goal prioritization.

A positive historical example:

A compelling illustration of the importance of prioritization can be witnessed in the narrative of Thomas Edison's journey to inventing the light bulb. Throughout his pursuit, Edison encountered countless challenges and encountered setbacks, yet he remained steadfast in his objective of developing a practical electric lighting solution. Undeterred by the obstacles, Edison devoted himself to relentless experimentation and innovation until he ultimately succeeded in accomplishing his goal.

A negative historical example:

A negative example of the failure to prioritize can be seen in the story of the ill-fated maiden voyage of the Titanic. The ship's designers and crew were so consumed by the ambition of building and operating the world's most luxurious cruise[74] ship that they neglected other crucial priorities, such as ensuring the safety and well-being of the passengers and crew. This lack of proper prioritization and emphasis on safety over luxury ultimately led to the tragic loss of the ship and numerous lives. Had the designers and the team placed a greater emphasis on prioritizing safety over the pursuit of extravagance, the disaster could potentially have been prevented.

---

[74] Right now we will be able to touch the archetype that led to the sinking of the Titanic. Officially, the Titanic was not considered a cruise ship, but an ocean liner. The difference is big, in fact: a cruise ship travels for fun, and an ocean-going ship has a schedule that it must comply with precisely because it is in charge of delivering the goods. But the fierce competition between the owners of the Titanic — White Star Line and its direct competitor — Cunard Cruise Line — led to the fact that the Titanic was built as a luxury ship at the design stage. The captain was aware of the ice situation, but, taking into account the schedule of very important passengers, he chose to take a chance. The ego of its passengers ruined the ship.

6. ANTICIPATE THE FUTURE

Predicting the future is an intriguing concept, but it's important to acknowledge that no method can guarantee accurate predictions. However, there are several techniques that can be used to make informed decisions based on statistical analysis, trends, intuition, and even divination tools such as tarot cards or coin flips.

Mathematical statistics and trend analysis involve analyzing past data and patterns to make predictions about future events. For example, stock traders use statistical analysis to predict market trends, while meteorologists rely on weather patterns to forecast future weather conditions.

The Copernican method offers a different approach by encouraging individuals to step back and view the situation from an outsider's perspective. This involves considering all possible outcomes based on different scenarios, helping to break down complex situations into simpler parts and provide a more objective viewpoint.

Intuition can also play a significant role in anticipating the future. Trusting one's gut feeling or instinct about a situation, even without a logical explanation, can provide valuable insights. Some people may use divination tools like tarot cards or coin flips to engage their intuition and gain further understanding. These tools emphasize the importance of making decisions swiftly, even when faced with information uncertainty.

It's important to note that none of these methods can guarantee absolute accuracy in predicting the future. Therefore, it's crucial to make decisions based on a combination of these methods and careful consideration of all available information. Additionally, recognizing the positive element of taking action, even with imperfect solutions, is important. Avoiding important decisions can lead to procrastination and missed opportunities. Ultimately, the key is to gather as much information as possible and use a combination of methods to make the best possible decision based on the available data.

7. WHO BENEFITS THE MOST FROM IMPLEMENTING BEST PRACTICES?

Trusting your own decisions is of utmost importance, particularly when faced with change or uncertainty. You possess a deep understanding of yourself that no one else can fully grasp. While it can be beneficial to seek advice and gather information from others, it is essential to trust your instincts and make decisions based on what resonates with your inner self.

There may be moments when you feel tempted to align with the opinions or choices of others, especially if they appear to be in the majority. However, it is crucial to recognize that their perspectives may not always align with your own. As the saying goes, "trillions of flies can make mistakes." Just because something is popular or widely accepted does not automatically make it the right path for you in a specific situation. Take the time to contemplate who and why a particular solution may serve your best interests. Sometimes, the truth lies unobscured, revealing the whole picture as if it were in the palm of your hand.

Trusting your own decisions also entails taking ownership of the choices you make and their consequences. Even if a decision does not unfold as expected, it presents an invaluable opportunity for growth and learning. Through reflection and analysis, you can deepen your self-awareness and cultivate unwavering confidence in your decision-making capabilities.

Nevertheless, it is vital to strike a harmonious balance between trusting your inner voice and remaining receptive to feedback and advice from trusted individuals. Exploring diverse perspectives and considering the opinions of respected advisors can provide valuable insights.

Ultimately, you alone bear the responsibility of living with the outcomes of your decisions. Thus, placing trust in yourself to navigate the complexities of your unique circumstances and honor your core values is paramount.

Areas in which it makes sense to be especially careful:

Choosing a career path

When it comes to choosing a career path, it's natural to seek advice and consider various factors such as job prospects, salaries, and personal skills. However, it's crucial to trust your instincts and make decisions based on what truly resonates with you. While friends and family may have their opinions, it's important to remember that their perspectives may not align with your own aspirations and passions. For instance, they may suggest a career in medicine, but if your heart lies in art, pursuing graphic design may lead to a more fulfilling and successful career. By following your own interests and staying true to yourself, you increase the likelihood of finding joy and success in your chosen profession.

Financial Decision-Making

When it comes to financial decision-making, gathering information and seeking advice are important steps to consider. However, at the end of the day, it's essential to trust your instincts and make decisions that align with your own financial goals and risk tolerance. While financial advisors can provide valuable insights, their recommendations may not always reflect your individual values or comfort level with risk. For instance, they may suggest investing in specific stocks, but if that doesn't resonate with your financial objectives or risk appetite, it's crucial to trust yourself and explore alternative options. By trusting your instincts and making decisions that align with your own financial situation and goals, you can navigate the stock market in a way that feels right for you.

Choosing a life partner

When it comes to choosing a life partner, seeking advice and considering various factors are important steps to take. However, ultimately, it's crucial to trust your instincts and make decisions based on what feels right to you. While friends and family can offer their perspectives, their opinions may not align with your own feelings and desires. For example, they might encourage you to stay in a relationship even if it doesn't bring you happiness. In such cases, it's essential to trust yourself and prioritize your own well-being when making decisions about your life partner. By trusting your instincts and following your own path, you can find a partner who truly aligns with your values and brings you fulfillment in the long run.

8. GIVE LESS REGARD TO THE IMPORTANT AND GRANT PROMINENCE TO THE SECONDARY

Enter the realm of "kaizen," the remarkable Japanese philosophy that champions the practice of dedicating significant attention to minor matters while intentionally reducing the importance of weightier ones. This profound approach calls for continuous improvement through incremental changes, permeating all aspects of life, from personal endeavors to the intricate workings of businesses. Pause and ponder: Are the seemingly unassailable rules truly impervious to challenge?

In the tapestry of kaizen, lies a transformative method — an intentional disbalance where the emphasis on seemingly trivial solutions or tasks surpasses that of their more significant counterparts. By dissecting grand aspirations into manageable fragments, our focus shifts towards incremental progress rather than being engulfed by the overwhelming scope of the bigger picture. This practice holds particular relevance in decisions with far-reaching consequences, fostering a balanced and deliberate approach.

By diverting attention to the minutiae, we unveil hidden realms of improvement that often elude our notice. In the realm of commerce, this entails scrutinizing processes and procedures, uncovering inefficiencies and identifying opportunities for optimization. Through the accumulation of small, calculated adjustments, the overall performance of an organization steadily ascends, inch by precious inch.

Yet, kaizen's impact extends beyond productivity, offering solace amidst life's tumult — a sanctuary of mindfulness and gratitude. By immersing ourselves in the present moment and embracing the often-overlooked details, we cultivate an unparalleled appreciation. Amidst the chaos and uncertainty, this mindful disbalance serves as a catalyst, grounding us and allowing us to unleash our full potential.

Embracing the essence of kaizen, by consciously allocating our energy to less significant matters and reducing the weight of those that appear grand, we unlock a powerful pathway to change the world for the better. Through this deliberate disbalance, we reclaim our time and maximize our impact. As we embark on this transformative journey of mindful choice, we have the capacity to create a profound ripple effect that not only enhances our lives but also resonates throughout the world we inhabit.

9. SUBJUGATE ACCIDENTS

The anchoring method is a cognitive distortion in which we rely too much on the piece of information that we consider to be the most important when making a decision. This initial information acts as an "anchor" that influences our subsequent judgments and decisions.

During the Second World War, the United States armed forces were faced with a difficult decision regarding the production of grenades, namely, to determine the maximum allowable rate of grenade defects. In other words, the military needed to set a limit on the number of defective grenades that could be produced before the entire batch was rejected.

Initially, a limit of 5% defects was set, based on the fact that it was the industry standard at the time. However, when a group of statisticians was brought in to evaluate the solution, they proposed a lower limit of 1% defects, based on the severity of the consequences if a faulty grenade failed. However, it was not possible to achieve the desired result even by placing an observer over the assembler, and then an observer over the observer. Moreover, the situation even became worse — transport ships with grenades and planes that transported boxes of grenades began to explode, since only grenades with hidden defects reached them, which appeared only during critical situations — pitching during a strong storm or shaking while falling into air pits.

In the end, however, the military was able to accept the recommendation of statisticians and set a limit of 1% defects. This decision was based on a thorough assessment of the risks and benefits involved, rather than just a speculative initial anchor of 5% defects. Interestingly, this was achieved only with the help of an interned American of Japanese descent , which invented such a holder in the conveyor mechanism, which did not allow a defective copy to be fixed on it, which immediately made the defects equal to zero, and American grenades were an absolutely reliable weapon. At the same time, no observers were required over the observers, and the error was now excluded.

The lesson from this example is that the anchoring method externally can help you solve the problem by creating many additional ones, but you can turn the situation around in such a way that its main limitation will be your complete victory through the absolute elimination of statistical errors. This means that you have put a random event at your service, removing all the possibilities of a negative scenario, but leaving all positive possibilities. Such an approach is justified in business, in the matter of choosing a partner, etc. — wherever there is an opportunity to set unacceptable conditions as an anchor that leaves unacceptable passengers on the shore.

Such a method is now better known by the name of *Setting the Margin of Error*, or *"Schmerzgrenze"*[75], and thus can be a useful approach in decision-making, as it draws a clear line beyond which certain options are not acceptable. By setting a limit, we can avoid a situation where the decisions we make cause us undue stress or harm, such as when dividing drugs according to their importance to critical health conditions. This can help us make more rational, deliberate decisions that align with our values and goals.

10. THE IMPACT OF A CAT ON STAGE

The presence of a cat strolling across the stage can easily divert our attention, drawing it away from the main event. When we attend a play, our focus naturally centers on the performers and their actions. However, when a cat unexpectedly makes an appearance, our attention inevitably shifts towards it, causing us to potentially miss significant details of the performance.

This occurrence, often referred to as the "attention-grabbing" effect, arises when an unforeseen stimulus

---

[75] *Schmerzgrenze* [ˈʃmɛʁts ˌɡʁɛntsə] — the smallest strength of the stimulus, at which the creature perceives the stimulus as pain.

captures our focus and hinders our ability to concentrate on the primary task at hand. In the context of a cat traversing the stage, the sudden and unexpected presence of an animal serves as a prime example of such a stimulus.

The significance of attention-grabbing in decision-making lies in its capacity to sway our judgments and choices. When our attention is diverted by unanticipated stimuli, crucial details may evade our notice, and we may fail to consider the full range of available information. Consequently, decision-making can become biased or incomplete.

For instance, within a business setting, imagine a manager reviewing a report concerning a potential investment opportunity. If an unforeseen interruption, such as a sudden phone call or email notification, captures the manager's attention, they may inadvertently overlook critical aspects of the document, leading to a suboptimal decision. The same principle applies when strong emotions, such as anger, envy, love, fear, or gratitude, become influential factors in decision-making.

To counteract the adverse effects of attention grabbing when making decisions, it is essential to remain cognizant of potential distractions and take proactive measures to mitigate their impact. This may involve cultivating a distraction-free environment, such as disabling notifications on electronic devices or scheduling decision-making activities during periods when external interruptions are less likely.

In essence, attention grabbing possesses the potential to exert a potent influence, diverting our focus from important events and shaping the course of decision-making. By recognizing and actively minimizing potential distractions, we empower ourselves to make informed and effective decisions that align with our goals and aspirations.

11. ACQUIRING A LONG-TERM PERSPECTIVE INSTEAD OF FIXATING ON IMMEDIATE REACTIONS

Recursion in communication involves continuously contemplating how others will respond to our actions and adapting our behavior accordingly. This process can sometimes lead to repetitive cycles of questioning, even when we already possess the answers.

To address this issue, it is beneficial to shift our focus towards developing a long-term strategy that places less emphasis on short-term circumstances and instead prioritizes the enduring values of our group. By adopting a statistical approach to predict behavior, we can diminish the impact of these cycles and make decisions founded on more comprehensive insights.

For instance, imagine attempting to anticipate a group's reaction to a new policy proposal. Rather than incessantly pondering their response to each individual step, we can examine the shared values and priorities of the group and strive to formulate policies that align with those fundamental principles. Additionally, employing data and statistical analysis enables us to assess past behaviors and trends, thus aiding us in predicting the group's future conduct.

This approach enables us to transcend the cycles of repetitive questioning and make well-informed strategic decisions rooted in data and analysis. It also directs our attention towards the group's long-term objectives and values, rather than fixating on the intricate details of each isolated action and ensuing reaction.

In essence, recursion in communication may occasionally engender cycles of questioning and an overemphasis on immediate actions, diverting our attention from long-term values and goals. By adopting a more statistical approach to predicting behavior and embracing long-term strategies, we can make informed decisions and mitigate the impact of these cycles.

12. GAINING INSIGHT FROM THE PAST: UNVEILING HIDDEN TRUTHS

Anticipating the past involves retrospectively examining historical events and extracting valuable insights that can inform present and future decision-making. While certain crucial elements of the past may intentionally remain obscured or unknown, their revelation can significantly impact the choices we make.

For instance, let's consider the decision to invest in a particular company. By analyzing its past financial performance, we can gauge its potential future trajectory. However, there may exist less apparent factors from the company's history, such as a record of ethical violations or mismanagement. Identifying and acknowledging these elements empowers us to make more informed investment decisions.

Furthermore, extrapolating the sequence of past events can serve as a valuable decision-making tool. Studying historical occurrences and understanding the causal relationships that led to specific outcomes enables us to anticipate similar future events. For instance, when contemplating the entry into a new market, we can examine how other companies have ventured into analogous markets and assess their level of success. This analysis provides us with valuable insights into the associated risks and rewards.

Moreover, uncovering hidden or unknown aspects of the past provides crucial context for addressing present challenges or anticipating future issues. For instance, when faced with a complex situation involving a client or employee, drawing from past experiences and comprehending the underlying motivations can enhance our ability to navigate the circumstances effectively.

Nevertheless, unearthing hidden truths from the past can be a challenging endeavor, often involving the deliberate concealment or erosion of information over time. To aid in this pursuit, several strategies can be employed:

1. *Conduct thorough research*: Commence the journey by delving into comprehensive research. This may involve examining historical records, conducting interviews with experts or witnesses, and gathering data from diverse sources. By amassing a wealth of information, a more comprehensive and accurate narrative can emerge.

2. *Identify patterns and inconsistencies*: As information accumulates, scrutinize it for patterns and inconsistencies that may indicate hidden truths. Conflicting accounts from multiple sources or subtle disparities can often unveil pivotal details that contribute to resolving critical situations. In forensic investigations, even the slightest deviation from anticipated results can lead to groundbreaking discoveries.

3. *Embrace diverse perspectives*: When analyzing historical events, it is crucial to embrace alternative viewpoints and sources. Expanding beyond official records and considering marginalized or underrepresented perspectives can offer a more nuanced understanding of past occurrences.

4. *Maintain open-mindedness*: Approaching the exploration of the past with a neutral mindset is vital. This entails challenging personal assumptions and biases, remaining receptive to information that may challenge existing beliefs or prior understandings of the situation.

In addition to these strategies, patience and persistence are key virtues in the pursuit of truth. Uncovering hidden truths demands time and a multidimensional approach, but the rewards include a deeper comprehension of historical events and the ability to make more informed decisions in the present.

Neglecting inconsistencies can lead to catastrophic consequences across various domains, including engineering, medicine, and finance. The Chernobyl disaster of 1986 serves as a harrowing reminder of the tragic fallout resulting from a disregard for safety regulations and the dismissal of inconsistencies in the reactor's design. The haunting parallels to the ill-fated Titanic disaster lie in the concealment of information by Soviet party officials.

In summary, anticipating the past serves as a valuable tool in decision-making, enabling us to extract insights from historical events and apply them to the present and future. By unearthing hidden truths, extrapolating past sequences of events, and grasping the contextual backdrop of current challenges, we can make more informed decisions rooted in a profound understanding of the situation at hand.

Evaluate the algorithms guiding your life choices at present and explore the potential for transformation.

Basic Actions:
self-esteem, self-discipline, sports, sleep, sex, self-satisfaction, small habits of self-care

I'**m an ardent devotee of Basic Action. For instance, I once embarked on hosting an economic radio program, which lasted a full two and a half years. To fulfill this endeavor, I needed to delve into the economic landscape each day. Preparing for the initial broadcast, which spanned just under an hour, demanded over four hours of my time. Devoting such an exorbitant amount of time to it was inconceivable. Nevertheless, adhering to a regular study of trends replaced the meticulous examination of corporate financial statements. With time, I discovered reference points that swiftly enabled me to assess the business's condition relative to the market and provide prompt forecasts — almost ten times faster. By analyzing overarching trends, I managed to curtail the time spent scrutinizing the market's economic conditions as a whole, along with the status of industries and individual companies. The economic impact of this strategy was staggering — my portfolio outperformed the vast majority of funds throughout the entire duration of the program, all the while consciously evading risky assets. The essence of Basic Action, in this instance, lay in the consistent effort to scrutinize the interplay of macro and micro market indicators.

SELF-DISCIPLINE

Self-discipline, a virtue of controlling actions, thoughts, and emotions to achieve specific goals, is a paramount trait. It enables individuals to remain steadfast, motivated, and productive amidst challenges and distractions.

New York's coat of arms bears the Latin phrase "*Excelsior*", signifying a perpetual ascent to higher heights. It embodies the notion of continuous improvement and the pursuit of excellence across all facets of life. Associated with self-discipline, it urges individuals to strive for betterment, unlock their true potential, and eschew mediocrity. If you are unacquainted with *Longfellow's*[76] poem *"Excelsior"*, seek its wisdom — it reveals a protagonist willingly risking their life without explanation, resolutely spurning worldly allurements, though a reason undoubtedly exists.

A kindred principle in the Japanese tradition is *kaizen*[77], meaning "continuous improvement" or "change for the better." This philosophy emphasizes the value of incremental, gradual enhancements over radical transformations. Kaizen endeavors to foster a culture of perpetual progress, urging all within an organization to regularly identify and implement small improvements.

Self-discipline forms a cornerstone of the kaizen philosophy, necessitating personal accountability and a commitment to continuous self-improvement. To embrace kaizen, individuals must possess the discipline to recognize areas in need of improvement, set goals, and consistently take action to attain them.

Cultivating self-discipline within the context of kaizen can be achieved through the establishment of a daily practice of introspection. Carve out moments each day for reflection on your objectives, the progress made, and areas ripe for refinement. Through regular contemplation of your work, you can identify behaviors impeding progress and gradually refine your approach.

In the realm of kaizen, self-discipline further entails prioritizing goals and remaining steadfastly focused on the

---

[76] *Henry Wadsworth Longfellow* (Eng. *Henry Wadsworth Longfellow*; February 27, 1807 – March 24, 1882) was an American poet and translator. He is the author of "The Song of Hiawatha" and other poems and poems.

[77] *Kaizen* (Jap. 改善, "improvement") is a concept, more often related to business activities, which constantly improves all functions and involves all employees, from the CEO to the assembly line workers. Kaizen also applies to processes such as procurement and logistics that cross organizational boundaries in the supply chain. It has been used in health care, psychotherapy, life coaching, government, and banking.

most critical tasks. It demands a heightened self-awareness and the ability to resist distractions and temptations that might derail progress.

Hence, self-discipline and the kaizen philosophy intertwine closely, compelling individuals to shoulder responsibility for their personal growth and strive for continuous, incremental advancement. By nurturing self-discipline and embracing kaizen, individuals can foster a culture of perpetual improvement and achieve long-term success in both their personal and professional lives.

Continuing the discourse on self-discipline, I am prepared to offer advice rooted in personal experience on how to break free from detrimental habits, unhealthy behaviors, toxic relationships, and any other hindrances. Personal experiences have taught me effective ways to quit smoking, employing wise and deliberate strategies to gradually displace nicotine. Therapists and medications can aid in overcoming addiction. I understand that these approaches require effort, introspection, and self-awareness regarding one's motivations. I propose my own mindfulness-based method. Ask yourself: Does this habit contribute to your happiness? If it does, then smoke, drink, or indulge in drugs until you perish in the alley if that's your definition of contentment. I have no qualms with that. I speak only of those addictions that impede your happiness. If such a habit or relationship obstructs your well-being, if your job makes you howl like a wolf, then here's my remedy: quit. Gracefully and calmly sever ties. No elaborate plans or family therapists are required. If it hampers your happiness — quit. Would you stay in a marriage that doesn't suit you for the sake of the children? Abandon it. To banish obsession, merely utter to yourself three times: "NO. NO. NO." Put an end to the toxic practice. It invariably works: I, along with countless others who made such decisions, ultimately did just that. **Give it a try**.

SELF-ESTEEM

In ancient times, self-esteem held profound significance, particularly in the illustrious realms of Rome and Greece. The Romans perceived it as a tether to social standing, intricately tied to one's ability to conduct oneself with propriety and dignity. Upholding honor and reputation became a creed, while those who faltered were deemed disrespectful, left to the capricious whims of the gods.

Greece, too, revered self-esteem as a cherished virtue. The measure of a person's worth rested upon the extent to which they esteemed themselves. It was an expectation that individuals should safeguard their honor, act honestly, and display respect towards others. Self-respect, it was believed, fostered divine favor and upheld societal harmony.

These ancient civilizations, Rome and Greece, left indelible imprints on the world's tapestry. The legal systems forged by the Romans served as beacons, inspiring and influencing numerous nations, establishing a framework of laws and justice. Rome's engineering marvels, including aqueducts, roads, and sewers, provided infrastructure that endures and evolves within modern societies. Moreover, the rich history of Rome lay the bedrock for Western culture, encompassing language, literature, art, and governance.

In Greece, philosophy and politics assumed the mantle of societal cornerstones. Figures like Plato and Aristotle loomed large as philosophical giants, while the establishment of democracy in Athens reverberated through the annals of time. Greek theaters, acting as platforms for social and political discourse, laid the foundation for modern dramatic arts. Additionally, the Olympic Games, originating in Greece in 776 BC, have ignited passion and camaraderie across nations, enduring through centuries.

Self-esteem, often overlooked, holds the key to a prosperous existence. Esteeming oneself signifies harboring a positive perception, embracing accomplishments, and recognizing the potential for further achievements. It emboldens individuals to pursue their passions and aspirations, fostering the audacity necessary to undertake risks en route to desired outcomes. Self-respect becomes the impetus for transformative changes that lead to ultimate triumph.

Esteeming oneself entails treating every aspect of life with reverence, encompassing physical and emotional well-being, relationships, work, and leisure. It encompasses nourishing oneself with wholesome sustenance, engaging in exercise, nurturing emotional welfare, establishing boundaries in relationships, and embracing accountability for one's actions. Self-esteem necessitates an honest appraisal of how one's deeds impact both short-term and long-term aspirations.

When self-respect permeates the core, decisions emerge predicated on long-term benefits, even when immediate outcomes appear unfavorable. It emboldens the capacity to decline when one discerns the incongruity between personal needs and external pressures, even if it inconveniences others. Criticism is received with tolerance, yet fortitude manifests when one detects injustice and must assert themselves.

Self-respect becomes an anchor amid tempestuous waters, enabling the identification of patterns within relationships and the understanding that certain situations yield no positive fruition. It instills the audacity to sever ties with relationships or circumstances that prove detrimental.

Ultimately, self-esteem paves the path to a triumphant existence, wherein values, principles, and beliefs remain unsullied. With confidence in one's decisions and the ability to make choices conducive to overall success, the realization of dreams and aspirations becomes all the more attainable.

SPORT

Sport, a vital endeavor for physical well-being, encompasses activities like swimming, running, cycling, and general fitness maintenance. It is worth noting that maintaining proper physical fitness positively influences overall health and overall well-being. Yet, the true essence lies beyond mere health. The act of compelling oneself to engage in sports, disregarding present desires, initiates a meditative state. In this moment, one transforms into a warrior, honing skills for battle. Will this warrior's spirit transform my world upon emerging from meditation[78], returning to the mundane? Is the essence of my existence a daily struggle?

I have a dear friend who battles a severe autoimmune disorder. During attacks, excruciating pain engulfs her muscles, rendering even the simplest movements Herculean tasks. In an act of defiance against her predicament, she diligently performs her daily routine during remission[79], until pain forces her into unconsciousness. I remember her resilience when my own laziness tempts me to skip class or shirk responsibilities.

Personally, I am drawn to applied sports, such as karate. This martial art allows one to exhibit strength of

---

[78] Without planning to deviate far from the subject of our discussion, I will only emphasize that studies by doctors show that up to 2/3 of the physical volume of brain activity is associated with movement control. This means that constant movement essentially trains 2/3 of your brain volume all the time. Try to put it into practice and watch the result.

[79] *Remission* (Latin. *remissio* "reduction, weakening") is the period of the course of a chronic disease, which is characterized by a significant weakening (incomplete remission) or disappearance (complete remission) of its symptoms (signs of the disease).

character, self-discipline, and the ability to control pain and injuries. As far as I know, there are no limitations to sports, save for the obvious ones: when engaging in gym activities, it is wise to thoroughly discuss the full range of exercises with a professional trainer beforehand, and then diligently and consistently execute the routines independently. In general, sports should become an integral part of one's weekly routine, but determining the frequency and intensity of training may require the guidance of a coach.

Now, who exactly is a coach within the context of our conversation? A coach, in this instance, is an individual who receives compensation for their guidance, provides written instructions, and assumes responsibility for any misguided advice, at the very least in monetary terms.

DREAM

Rest, an elemental pursuit, is indispensable for one's overall health and well-being. A restful slumber maintains emotional equilibrium while affording the body and mind ample time for rejuvenation and restoration. Ideally, individuals should strive for 7 to 9 hours of sleep each night. Those who achieve blissful repose merit congratulations. However, those plagued by sleep troubles shall find this source of knowledge elusive. Let me explain why: prophetic dreams grace individuals during the middle third of the nocturnal darkness. Hence, one must divide the dark period into three equal parts. To achieve this, consult the precise times of sunset and sunrise. For instance, if sunset occurs at 9 p.m. and sunrise at 6 a.m., you have nine hours of darkness at your disposal. The window for prophetic dreams unveils itself solely during the middle third, which in this case falls between midnight and 3 a.m. It is vital to accord utmost attention to this information and promptly record the dream while still nestled in bed. To facilitate this, keep writing materials and a light source within reach, enabling you to transcribe every detail. Sleep should occupy the majority of one's day.

Pragmatic methods for tranquility warrant judicious employment. Personally, I employ lavender pads[80] that rest beside my pillow, and the aroma of balsamic[81] essence. Four drops of balsamic in a glass of water before bed yield skin improvement, stabilized blood sugar, enhanced digestion, lowered cholesterol, and improved blood circulation, among other benefits.

Guard your sleep with utmost care!

SEX

Let's clarify: we speak of the act of intercourse, specifically between a man and a woman. In the Kabbalistic tradition, intercourse symbolizes the unity of the world,

---

[80] The ancients associated lavender with the third eye chakra, attributes of which include intuition, imagination, visualization, and concentration.

[81] The history of the original balsamic vinegar has its origins in ancient Rome: known by the Latin name sapa, it was obtained by boiling grape juice (other historians attribute this discovery much earlier, about 3000 B.C. to Egyptian culture). Over the centuries, the idea of aging saps in wooden barrels has become popular in areas of present day Emilia–Romagna; this is demonstrated by an excerpt from the eleventh-century poem "Vita Matildis", which tells of an Italian prince who brings a precious condiment as a gift to Henry III of France. Balsamic vinegar grew in popularity over time, and in some Renaissance texts it is referred to as "Duke's vinegar". The seasoning takes the name to which we became accustomed in 1747: in fact, the adjective "balsamic" first appears in the inventory of the noble Este family. The word comes from Latin balsamum, a term used to describe a beneficial plant: this is probably due to the fact that in the 18th century balsamic vinegar was used to treat the plague; Until now, many doctors claim that this seasoning is a powerful ally in preventing circulatory, digestive and diabetes problems.

akin to the fusion of *yin* and *yang*[82] in Eastern teachings. It presents a grand opportunity to ascend into the realm of ideas, Bina[83]. The arrangement is such that a man lacks direct access to the world of ideas, although his nature impels him towards an intense focus on his primary task. On the other hand, a woman possesses a more effortless entrance into the world of ideas, albeit with an unfocused perception of the information. When a man and a woman come together, they form an analog code that allows for the exchange of information. The man imparts elements of his

---

[82] *Yin* and *yang* (Chinese trad. 陰陽, control. 阴阳, pinyin yīn yáng; Yap. 陰陽 ying-yo :) is a stage of initial cosmogenesis in the view of Chinese philosophy, the acquisition of the greatest separation of two opposite properties. Graphically indicated by the appearance of two opposites of two colors — white and black.

[83] *Biná* (Dr.-Heb. בינה; bīnāh; "Mind"; "Mind"; "Thought") — in the teachings of Kabbalah on the origin of the worlds, the third of the 10 objective emanations (direct rays of divine light) of the universe — the so-called "Sefirot" or "Sefirot" (plural from "sefir"), also "numbers" or "spheres" — the first radiations of the Divine Essence, which together form the cosmos.

Conceived as members of a whole, the Sephiroth form the form of a perfect being, the primordial man (Adam–Kadmon). For greater clarity, Kabbalists indicate the correspondence of individual sefirot with the outer parts of the human body: for example, Binah and Hochma are the two eyes (eyelids) of Adam–Kadmon.

The first Sefirot — Light (*Keter*), Wisdom (*Hochmá*) and Reason (*Biná*) — the "forehead" and "eyelids" of Adam–Kadmon — are the most important. In their absolute realm of spirit, there is no place for any duality. They are followed by seven minor Sefirot. The bifurcation of the emanation into positive and negative begins only in the third sefirah ("Reason"). The Sefirot, who have their origin in the third sefir, form the basis for the entire lower material world. The contrasts and contradictions that prevail in the world can first manifest themselves only in the field of the third sephirah.

Two parallel principles (*Hochmá* and *Biná*), outwardly as if opposite, are in fact completely inseparable from each other: Hochmá ("Wisdom") is a masculine, active principle; Biná ("Mind") is feminine, passive. Together, Hochma and Bina give *Daat* ("Knowledge")

experience, often through semantic links akin to the *kata*[84] in martial arts, while the woman grants the man a pass into the world of ideas. According to the Kabbalistic tradition, the most effective way to access this realm is through direct access, which a man consciously or unconsciously attains after the woman experiences orgasm, followed by his own ejaculation. This often leads to finding solutions to professional challenges. I have personally observed and received confirmation of this phenomenon countless times. It is no wonder that our ancestors equated sex and money as a transfer of energy. However, in this book, we will focus on the energy of money. Thus, sex becomes another fundamental action, manifesting in various forms. For many, it brings satisfaction, pleasurable stimulation, and deepens intimate connections with others. Countless volumes have been written on sex, but I only wish to address the aspect of male *libido*[85] — how to ensure that

---

[84] *Káta* (Jap. 型 or 形) is a formalized sequence of movements connected by the principles of conducting a duel with an imaginary opponent or group of opponents. In fact, it is the quintessence of the technique of a particular style of martial arts. Similar to taol in wushu and thil in taekwondo. In the school of modern karate Joshindo, the concept of kata is additionally introduced into the definition of kata as a reference example of karate technique for imitation and study.

The principle of learning a martial art based on kata is that by repeating kata many thousands of times, a martial art practitioner accustoms his body to a certain kind of movement, bringing them to an unconscious level. Thus, getting into a combat situation, the body works "on its own" on the basis of reflexes embedded by repeated repetition of kata. It is also believed that kata have a meditative effect.

[85] *Libidó* (Latin lī́bīdo — lust, desire, passion, desire) is one of the basic concepts of psychoanalysis developed by Sigmund Freud to describe a variety of manifestations of sexuality. It denotes some specific energy underlying sexual desire.

Sigmund Freud equated libido with Plato's eros and defined it as the energy of attraction — the basis of sexual love, as well as any other (for example, love for parents and children). According to Freud, in the narrow sense, libido means psychic energy that can only be

sexual relations continue to evolve and that male desire does not diminish over time? On one hand, this is indeed a matter of colossal importance, given that countless marriages crumble due to sexual imbalance. However, it does not solve all issues within a marriage. The practice I describe below aids a man's libido but does not assist a woman in cultivating love for her partner. To comprehend how to continually enhance libido, we must understand the principles underlying the interaction between man and woman through sexual energy. To illustrate, envision the man as an open channel of energy, while the woman serves as the vessel that contains and stores this energy. This analogy accurately portrays their dynamic.

> Within the simplest and most effective Kabbalistic practices I have encountered for sustaining male libido in relationships, a seven-step system exists:

1. Before intercourse, it is wise for a man to prepare his partner through caresses, much like we ready a battery to receive energy. Of course, discovering her erogenous zones through benevolent conversation that nurtures trust makes it easier.

2. During intercourse, the partners should exchange breath at least three times. This entails synchronizing their breaths, inhaling and exhaling alternately, often during kissing. No lengthy explanations are needed; this can be achieved effortlessly.

---

discharged by sexual gratification, and in a broader sense, libido is the energy of the instincts of life, any psychic energy, underlying the desire for creation, love and harmony. The term "libido" was used by Freud to explain the causes of mental disorders, neurosis, as well as to describe the course of human mental development. In the transformation of libido (the so-called sublimation), Freud saw sources of creative energy. In modern sexology, the term libido is used, as a rule, in the meaning of "sexual desire"

3. There are no restrictions on the postures adopted during intercourse. However, considering the transfer of energy and its storage within a vessel, there is a central requirement: the woman should experience orgasm before the man. If both partners regard this as a fundamental prerequisite, it can always be accomplished. Ironically, not all women are prepared to reach orgasm with every sexual encounter, but it is crucial for fostering male libido growth.

4. The man should theoretically strive to ejaculate promptly after the woman, as the sooner it happens following her orgasm, the more effectively he retains his energy within her.

5. The man's climax should occur solely within the vessel, adhering to this condition's utmost importance. Any form of contraception is allowed and encouraged.

6. The choice of the most energetically suitable location for the man's climax is permissible, including within the vagina among the three possibilities.

7. After climax, the man should tenderly soothe the woman with unhurried caresses, ideally allowing her to quickly fall asleep — a truly ideal scenario.

Additionally, I am aware of a method to increase a woman's interest in intercourse and cultivate prearousal. Sleep also plays a vital role in sexual health. There exists a meditation practice that prepares a woman for proper intercourse.

While lying down or sitting, place one hand on your stomach and the other on your chest. Inhale deeply, feeling the expansion of your chest, then your abdomen. As you exhale, focus on the tension in the muscles of your lower abdomen and pelvis. Concentrate your energy in the pelvic region as you breathe out.

Visualize love and energy radiating from your heart throughout your entire body as you exhale. Allow them to envelop you, accumulating slowly.

As the energy builds, try breathing heavily through your mouth a few times. Many women find that this alone can arouse them. Imagine your breath descending down your chest as you inhale and rising up your back as you exhale.

Once you feel comfortable with this rhythm, begin gently contracting your pelvic muscles with each exhale to direct blood flow to your vagina.

Repeat this exercise regularly to develop control over your body and muscles. With frequent practice, you will feel more at ease, entering a state of arousal that expedites the process during sexual encounters.

SELF-SATISFACTION

Achieving a medical equilibrium in the realm of self-satisfaction involves two crucial components. Firstly, one must grasp the positive and negative aspects of this practice. Masturbation often provides a swift surge of energy and subsequent tranquility, aiding in alleviating heightened irritability and enhancing the quality of sleep. However, it is imperative to recognize that excessive orgasmic release through masturbation can lead to irritation and even depression. For instance, attempting two sessions of self-gratification before bedtime will likely hinder the ability to fall asleep.

Secondly, attaining medical balance in masturbation necessitates taking measures to circumvent contextual predicaments. The primary concern lies in developing disinterest in engaging with a woman. From a medical standpoint, this can manifest as erectile dysfunction in the worst-case scenario or a diminished inclination for new connections with women, as the emotional toll involved can prove formidable. Repeatedly exposing oneself to the risk of rejection, often less than amicable, poses a challenge, although there is certainly no fault in a man expressing his desire for a relationship with a woman. One way to maintain emotional control without sacrificing libido through self-satisfaction is to establish a sense of organization. For instance, predefining the frequency of this act, such as once a week on specific days, can prove beneficial. Another crucial aspect involves abstaining entirely from stimulating videos or photos, allowing your attraction levels to be self-sustained at a sufficient level for the progressive development of a relationship with a woman. Striving to maintain a relationship devoid of any anxiety becomes paramount, and this can be achieved by upholding a balance of basic actions.

COMPLETION BIAS AND ROUTINE

The human tendency to prioritize urgent tasks over important ones and the concept of completion bias hold significant importance. Completion bias refers to the brain's innate desire for closure and the pleasure it brings. The act of completing simple, immediate tasks releases dopamine, which enhances attention, memory, and motivation, creating a positive feedback loop. Starting the day by making your bed, for example, can provide a sense of completion and affirming positivity, increasing the likelihood of a successful day. This feeling of completion is linked to a chemical project at the core of our existence.

A recent study involving over 500 employees from various industries explored the impact of task completion on job satisfaction, motivation, and productivity. The study found that employees who began their day by completing a few short, mundane tasks and checking them off their lists experienced higher levels of job satisfaction, motivation, and overall productivity throughout the week.

Additionally, completing small tasks frees up cognitive resources, allowing individuals to focus better on other activities.

However, it is crucial to exercise caution against dedicating excessive time to mundane tasks at the expense of important projects. Some studies suggest that incomplete tasks occupy the mind, making it difficult to concentrate on other activities. In emergency departments, physicians may tend to prioritize easy tasks when faced with a higher workload, potentially leading to longer wait times for patients with more severe conditions.

To strike a balance and avoid succumbing to completion bias, several strategies can be considered. Firstly, individuals should identify their top three to five priorities and allocate sufficient time to them. Priorities should be explicitly defined and adjusted as needed. Secondly, individuals can harness completion bias by tackling a few mundane tasks at the beginning of the workday, priming their minds for more significant activities thereafter.

By adopting a thoughtful approach to structuring their work routines, individuals can leverage completion bias to enhance productivity and effectiveness.

SMALL HABITS OF SELF-CARE

The significance of self-care is widely acknowledged, and integrating meaningful actions into daily routines can have a profound impact on well-being. Self-care goes beyond occasional indulgences or extravagant experiences; it encompasses simple yet impactful actions that contribute to overall well-being. The key lies in identifying activities that activate the brain's four "happiness chemicals": dopamine, oxytocin, serotonin, and endorphins.

Dopamine, associated with motivation and reward, is released when tasks are completed, goals are achieved, or enjoyable food is savored. Oxytocin is generated through experiences of love, friendship, and quality time spent with loved ones. Serotonin, connected to pride and recognition, is triggered by acts of kindness and acknowledging positive accomplishments. Endorphins are released during physical activity, persistence, and intense moments.

Engaging in self-care activities that stimulate these chemicals is crucial for building resilience and preventing burnout. Instead of attempting drastic overhauls, incorporating small daily habits of self-care is recommended. Even simple forms of physical activity, such as walking or taking the stairs, can significantly boost mood and well-being.

To maximize the benefits, consider incorporating elements that trigger the release of happiness chemicals. Spending time with loved ones, particularly in natural settings, enhances the effects of physical activity. Performing acts of kindness, engaging in playful activities with children, listening to enjoyable music, and celebrating progress through visual cues like ticking off completed tasks in a calendar all contribute to a more fulfilling self-care routine.

The key takeaway is that self-care should be an integral part of daily life, not just a response to stressful situations. By identifying a simple activity and committing to it consistently, individuals can experience the profound impact of tiny habits on their overall well-being.

The COVID-19 pandemic has presented unique challenges that can negatively impact mental health, but it has also provided valuable insights. Prioritizing our well-being and that of our families during these trying times is essential. Our brains play a vital role in maintaining and increasing our "feel-good" chemicals, which can help us cope with stress and promote overall mental health.

When we experience stress, the brain releases cortisol, a chemical that acts as an alarm, signaling potential threats to the body. In the past, cortisol helped our ancestors survive in dangerous situations. However, in the context of social isolation and uncertainty, cortisol can be triggered more frequently, leading to increased stress levels. To counteract this, we can activate our "feel-good" chemicals through specific behaviors and habits.

> Here are the four main "feel-good" chemicals and suggestions on how to stimulate them:

Dopamine

Dopamine is associated with feelings of excitement and joy[86]. It is released when we anticipate happiness or work towards achieving goals. To increase dopamine levels:

- Schedule daily activities to look forward to.
- Plan timed activities with a goal to beat the clock.
- Engage in family competitions and games.
- Consume foods rich in tyrosine, such as eggs, chicken, almonds, and bananas.
- Use checklists to experience a sense of achievement when completing tasks.
- Plan enjoyable activities for when chores and responsibilities are finished.
- Plan future events and holidays to look forward to.

Oxytocin

Oxytocin is known as the "love hormone" and is released when we feel close to others. It promotes social bonding and trust. To boost oxytocin levels:

---

[86] Happiness is an emotion in which one experiences feelings ranging from contentment and satisfaction to bliss and intense pleasure. Joy is a stronger, less common feeling than happiness. Witnessing or achieving selflessness to the point of personal sacrifice frequently triggers this emotion.

- Increase physical affection, such as cuddling, massages, or snuggling on the couch.
- Connect with friends and family online through video calls or messaging platforms.
- Record videos of various activities to share with loved ones.
- Write letters or make cards to send to friends and family.
- Engage in virtual games or activities with others.
- Plan activities that benefit the community, such as helping the homeless or supporting neighbors.

Endorphins

Endorphins are released in response to physical exercise or anticipated pain. They act as natural painkillers and contribute to the "runner's high" after a workout. To stimulate endorphin release:

- Engage in activities that induce laughter and joy.
- Incorporate daily physical exercise, such as walking, running, or biking.
- Use at-home workout equipment or follow online exercise videos.
- Play sports in the backyard or isolated areas.
- Utilize apps or online platforms offering yoga, cardio, or other exercise classes.
- Participate in movement videos or programs specifically designed to boost endorphin release.

Serotonin

Serotonin is linked to feelings of respect, appreciation, and reward. It is released when we perceive ourselves as strong or assertive. To increase serotonin levels:

- Consume foods rich in tryptophan, such as poultry, spinach, nuts/seeds, salmon, soy, and milk.
- Spend quality one-on-one time with each family member.

- Assign leadership roles within the family, allowing individuals to make decisions and contribute.
- Acknowledge and appreciate the positive actions and qualities of family members.
- Provide praise and compliments for achievements or efforts.
- Encourage conversations about positive experiences and observations within the family.
- Maintain regular meal and snack schedules to ensure proper nutrition.

It's important to be aware of the activation of cortisol and its effects on our outlook. When cortisol is triggered, we tend to focus on potential threats. Remind yourself that you have control over your brain and actively engage in activities that stimulate your "feel-good" chemicals.

By incorporating these strategies into your daily life, you can nurture your mental well-being and promote resilience during the challenging times of the pandemic. Remember that self-care and taking care of your mental health are essential, and by focusing on activating your "feel-good" chemicals, you can support your overall well-being and resilience.

INDIRECT METHODS
TO INCREASE SEROTONIN LEVELS

Nonpharmacologic approaches to boosting serotonin levels in the brain are worth exploring. Serotonin, a neurotransmitter crucial in mood regulation and often associated with depression and other mental health conditions, can be influenced through various means. While there are numerous approaches, the following ones are particularly noteworthy:

1. *Altering thought and mood*: Self-induced changes in mood can potentially impact serotonin synthesis in the brain. The interaction between serotonin synthesis and mood may be two-way, with

serotonin influencing mood and mood influencing serotonin. Psychotherapy and other techniques that improve mood may have the potential to increase serotonin synthesis.

2. *Exposure to bright light*: Bright light exposure, such as sunlight, has a positive impact on mood. It is a standard treatment for seasonal affective disorder, and some studies suggest it may also be effective in nonseasonal depression. Research indicates a correlation between serotonin synthesis and the hours of sunlight, indicating that bright light exposure may increase serotonin levels.

3. *Adopting a pet*: Owning a pet offers numerous health benefits, both physical and emotional. Research has shown that having a pet can reduce stress levels by lowering the stress hormone cortisol and increasing levels of the feel-good hormone oxytocin. The interaction between individuals and their pets has been found to be particularly effective in reducing symptoms of post-traumatic stress disorder, with a significant number of patients reporting a reduction in symptoms and a decrease in medication usage.

Let's delve further into the last option. Interacting with pets, such as petting them or engaging in activities with them, can help lower blood pressure, providing relaxation and a sense of transcendence. Additionally, owning a dog can increase physical activity levels, as dog owners often have to walk their pets regularly, ensuring a daily dose of exercise. This, in turn, can boost heart health and help prevent cardiovascular disease.

Pets also provide companionship and can help alleviate feelings of loneliness and depression. They offer a sense of purpose to their owners. By caring for and nurturing a pet, individuals can experience a profound impact on their overall well-being.

It's important to explore different avenues and find what works best for you when it comes to increasing serotonin levels indirectly. These methods can complement other treatment approaches and contribute to a more balanced and fulfilling life. Remember to prioritize your mental health and seek support when needed.

Ponder these musings with utmost deliberation.

Chapter VI

Overcoming
Natural Obstacles

Overcoming natural obstacles has always been an intrinsic part of our existence. Nature presents us with an array of challenges, be it rugged terrains, imposing boulders, or even the elements themselves — droughts, floods, strong winds, and tempests that test the resilience of all that resides on this earth. There are also subtler obstacles that manifest in the form of parasites, beetles, and other organisms that disrupt the natural order. Tornadoes, storms, and fires complete the tapestry of natural hindrances.

In truth, we could inhabit a safer world, shielded from the concerns of inclement weather and other adversities. Our villages, towns, and cities provide refuge, at least from these particular hazards. Yet, we continue to encounter a multitude of obstacles born of human influence — internal struggles within our own minds, biases and phobias, and the everpresent temptation to compare ourselves to our neighbors. Interpersonal challenges arise as others seek to pilfer our possessions, ideas, or time. Societal obstacles emerge in the form of political turmoil, financial crises instigated by external forces, competition, or the daily toil required to make ends meet, to achieve success, recognition, and happiness in our relationships.

When we observe nature, we witness how it confronts obstacles with unwavering composure. Nature does not succumb to panic, nor does it complain or stress. Instead, it instinctively seeks new paths whenever an obstacle arises. Like a blade of grass that gradually pierces through asphalt, nature perseveres through slow, steadfast effort over time. It does not lament; it simply follows its innate purpose — growing and reaching towards the sunlight, finding every crevice and crack through which to flourish.

Similar examples can be found throughout the natural world. The forest rebounds after fires, droughts, storms, and floods — given enough time. It does so through gradual, methodical, and unwavering endeavor.

Let us emulate nature's approach to overcoming obstacles in our own lives. Let us closely examine the conditions in which nature thrives. Perhaps, we can discern a fundamental difference — the reaction. There is an ancient Kabbalistic tale about a tiger, an elephant, and a jackal, each representing one of the primary animal reactions prevalent in nature — aggression, fear, and stupor, respectively. When an elephant encounters a tiger, it could attack, but it refrains because elephants do not prey on tigers, and tigers do not hunt elephants. Nature dictates the absence of reaction. Likewise, when a tiger encounters a jackal, the jackal remains composed since there is no point in the tiger attacking — a hungry tiger expending energy on a jackal would prove futile. Again, nature dictates the absence of reaction. Lastly, when two tigers meet, theoretically competitors for hunting grounds and mates, a *casus belli*[87], cause for conflict arises. Yet, nature distinguishes the context: if it is not the mating season and there is an abundance of hunting grounds, the two tigers need not engage in perpetual stalemate. Once again, nature chooses the path of non-reaction.

In essence, nature teaches us not to overreact — to panic, stress, or despair in response to external circumstances, which, for the purpose of this discourse, we shall refer to as triggers. Instead, view these triggers as events that have already occurred, realizing that no amount of despair will alter their presence. They are part of your life now, and that is the new norm. Observe the situation, analyze it, comprehend what lies before you, and remain calm while fostering unwavering faith in yourself. Base your steadfastness on the following objective actions:

---

[87] *Casus belli* (from Latin. *casus belli* "a pretext for war"; pl. *casus belli*) is an action or event that either provokes or is used to justify war.

- Focus on finding solutions rather than fixating on problems. As in nature, any obstacle can be overcome with sufficient time, employing gradual and consistent effort. Pay heed to the gradual aspect — do not rush or attempt to conquer everything at once, risking exhaustion or jeopardizing your well-being. Nature does not make hasty and feverish decisions; instead, it achieves resolution over time.
- Persevere and explore alternative paths. Your initial attempts may not yield the desired results, but each failure eliminates more options. Without further exploration, you may not find the way around the obstacle. The first, second, or even third attempt may fall short, but eventually, one will succeed.

HOW TO OVERCOME OBSTACLES IN BUSINESS

Whether you forge a career or embark on the entrepreneurial path, you will inevitably encounter numerous obstacles — legal hurdles, accounting challenges, competition, financial setbacks, and more. Yet, you can surmount them all by adhering to the principles of nature. Refrain from agitation or panic; instead, observe, seek solutions, and navigate around the obstacles. Once you identify a solution, diligently work on it, gradually improving the situation until the problem becomes obsolete. In project management, this principle is manifested through the division of large tasks into smaller, manageable components.

Express gratitude for your obstacles. In many instances, an obstacle represents an opportunity for learning and growth. Established businesses often arise from addressing imperfections in the market. They do not emerge overnight as brilliant ideas from brilliant minds; rather, they gradually evolve by finding solutions to obstacles. Often, they diverge significantly from their original concept — much like the Benedictine monks who unwittingly created champagne. Originally, they had no intention of producing it. Champagne came into existence through happenstance[88]. Winemakers initially viewed the effervescence in champagne as a flaw. However, recognizing the distinctiveness of the drink, they resolved to refine it.

The residents of the Champagne region longed for the same level of popularity enjoyed by beverages originating from Burgundy. Determined to achieve such acclaim, they faced unforeseen obstacles due to their cooler climate — high acidity and low sugar content in the fruits, coupled with inadequate grape ripening. As a result, their wines lacked vibrancy and density, a stark contrast to the wines of Burgundy. However, the same climate slowed fermentation and reduced its rate, providing an opportunity for meticulous refinement.

---

[88] This is, in fact, an important and interesting aspect: *Dom Pérignon* — the same monk who in 1668 took the post of manager of the Benedictine abbey, was not the inventor of traditional champagne, since the effect of champagne wines was already known, but was considered a defect due to the very sharp taste, which, however, the monk coped with and it was he who accelerated the process of obtaining the famous alcohol.

The monk considered his main task to rid the wine of bubbles, and therefore created particularly harsh methods to reduce the possibility of this drawback:

- pruned the vine so that its length was no more than 1 meter;
- picked grapes only at dawn, while the fruits had not yet had time to warm up in the sun;
- used only donkeys and mules to transport bunches, believing that they were calmer than horses, which means that the grapes would arrive whole;
- I reviewed each grape and disposed of those that were slightly dented or completely crushed.

HOW TO OVERCOME OBSTACLES
IN EXERCISE AND NUTRITION

I perceive value in modifying our bodies through sustained efforts that contribute to long term health and a fit physique. To accomplish this, gradual and consistent exertion over time is paramount. Allow me to share the most vital insight I have about proper nutrition: eat to satiate hunger, not to compensate for failure or alleviate stress. Disregard societal pressures to indulge excessively at social or religious gatherings.

Rapid weight-loss diets yield no lasting results. Instead, strive to change your eating habits and overall lifestyle in a manner that can be sustained over the long term — perhaps a lifetime. Only through prolonged, continuous effort and positive changes will you witness the desired outcomes — weight loss, muscle gain and retention, improved agility, and overall well-being. Nevertheless, from a practical standpoint, I have found a particular practice to be effective. When I need to shed excess weight, I abstain from eating for five consecutive days, allowing only water or tea. This period of abstinence enables me to evaluate my current dietary habits and identify any mistakes. By the third day, hunger dissipates, and by the sixth day, I am free from its grip. On the sixth day, I allow myself a light vegetable salad. After fasting, nutritious food seems like a gift from heaven. The following day, I can eat whatever I desire, but the memory of the satisfaction derived from consuming wholesome food lingers. This empowers me to make dietary changes tailored to my specific needs. Gradually, I learn to consume less food overall and transition to a measured diet that leaves me slightly hungry after meals. All of this occurs with minimal stress, enabling me to make conscious food choices.

However, moderation is key. Engaging in excessively rigorous exercises daily will not yield the desired outcomes.

There is such a thing as overexertion, where muscles lack sufficient time to recover and grow. Nevertheless, I do not fear daily workouts; instead, I distribute their intensity throughout the week. Exercising wisely, allowing ample rest, and, above all, maintaining consistency over extended periods facilitate lasting transformations.

It is essential to differentiate between the manifestation of change and the need for meaningful action. For instance, I adopt a stricter diet when preparing for competitions or when my weight deviates from the desired range. I begin with the aforementioned five-day fast, followed by intermittent fasting (eating every other day for ten days) upon completion. Proper nutrition imparts a valuable lesson, one taught by nature itself — eat only when hungry.

Lastly, I emphasize the importance of making your own decisions concerning your health, appearance, and muscle mass. In Japanese tradition, the concept of "ubaitori"[89] encapsulates this idea, symbolized by four iconic spring-blooming trees — the cherry tree, plum tree, peach tree, and apricot tree — each unique yet thriving side by side. This concept harmonizes seamlessly with the notion of "ikigai," emphasizing the significance of cherishing your distinct qualities, traits, and abilities.

Reflect upon these thoughts attentively.

---

[89] *Ubaitori* is one of the idioms of Yojijukugo (Japanese: Yojijukugo). 四字熟語) is a Japanese lexeme consisting of four kanji. In a broad sense, refers to Japanese compound words consisting of four kanji characters. However, in a narrow or strict sense, the term refers only to compounds of the four kanji that have a special (idiomatic) meaning that cannot be deduced from the meanings of their constituent components.

Chapter VII

Confronting Internal Enemies

I nternal resistance, that battle within oneself, remains an ever-present obstacle. It hinders productivity and obstructs progress, both individually and within organizations. This resistance can take various forms — a collective attitude or individual opposition rooted in fear, habit, or attachment to the familiar. Often, it materializes as a negative mindset and a staunch resistance to new ideas. In the realm of productivity, internal resistance slows the adoption of novel strategies or technologies, impeding the achievement of desired goals.

Have you ever experienced the reluctance to fulfill a promise due to inner doubts? On the day I penned this segment, a financial broker called regarding a significant trade. However, I declined to proceed with this particular broker, convinced of their lack of belief in success. Instead, I pursued the deal with an advisor who embraced our vision to complete a long-standing construction project using modern energy-saving technologies and renewable energy sources, thereby making housing more affordable. This approach qualified us for tax benefits and infused the deal with emotional appeal. Although the project incurred millions of dollars in additional costs, the advisor focused on the future rather than dwelling on the past — and the funds were secured.

REPTILE BRAIN

Allow me to clarify — I do not refer to an actual reptile. The term "*lizard brain*" describes the emotional processing part of our brain[90], scientifically known as the

---

[90] *The Triune Brain* is a model of forebrain evolution and vertebrate behavior proposed by American physician and neuroscientist Paul D. MacLean in the 1960s. The triune brain consists of a complex of reptiles (basal ganglia), a complex of paleomalec (limbic system) and a complex of mammals (neocortex), each of which is considered as independently conscious and as structures successively added to the forebrain in the course of evolution. According to the model,

amygdala. This primal region once served as our defense mechanism, instinctively reacting to threats without conscious thought, safeguarding us from harm. However, in our rapidly evolving civilization, this reptilian brain, resistant to change, has become a formidable barrier to growth and progress.

Every task we undertake carries an inherent risk, most notably the risk of failure. These challenges involve elements of change that activate the reptile brain. It constructs a barrier in the form of doubts about the chosen solution, apprehensions about the consequences of failure, and more. These questions form the foundation of internal resistance, impeding our innate desire for progress. When resistance prevails, we postpone tasks or even sabotage our own efforts to our detriment.

FLIGHT OR FIGHT?

The lizard brain also triggers the fight-or-flight response. When faced with a challenge, this brain reacts with either paralysis, confrontation, or evasion.

"Fighting" within our limited context entails combating the internal resistance spawned by our lizard brain and dealing with the tasks at hand. "Flight," on the other hand, represents avoidance. When we choose to "escape," we yield to internal resistance, shying away from tasks that involve risk and potential change. Paralysis, the most severe stage of internal resistance, renders us defenseless against external circumstances.

All of this, undoubtedly intriguing, poses a practical question: How do we confront the panic that engulfs us when faced with a task? Below, I present methods for

---

the basal ganglia are responsible for our primary instincts, the limbic system is responsible for our emotions, and the neocortex is responsible for objective or rational thoughts.

achieving a specific outcome — loosening the iron grip of your ego.

Ignore

One might be tempted to disregard the reptile brain's moans and proceed confidently with our brilliant plans. However, this approach proves unwise, as it only leads to accumulating risks that will inevitably manifest, regardless of our wishes.

Agree

So, what alternative exists to abandoning the risky strategies feared by your reptile brain? The answer lies in negotiating with this primal brain. We must implement various mechanisms to protect against and mitigate risks, mechanisms that the reptile brain will find credible. As previously discussed, risks can be segmented, addressing them as they arise or escalate.

Anticipate surprises

A more advanced approach involves incorporating failure into the framework. Since failure is an inherent part of the process, allocating room for 3–4 failures within your enterprise budget while developing an effective business algorithm allows you to adjust your actions as you work, minimizing the risk of abandoning the project entirely.

Give it your name

I once found myself in a discussion with an unfamiliar individual about a deal. However, as the conversation progressed, it became clear that my decision would be unprofitable in the long run. I had driven an hour and a half to meet them, but that wasn't reason enough to waste several weeks on futile negotiations or, worse, make an undesirable deal — an unfortunate experience I had encountered before. So, I promptly expressed my lack of confidence in the prospect, explaining the reasons. We shared a laugh and decided to simply enjoy a cup of tea.

During our conversation, we unexpectedly stumbled upon two assets that seemed complementary. Excitement engulfed me as I realized the incredible profitability of a mutually beneficial deal. However, an unexpected problem arose — I began coughing uncontrollably, unable to discuss the matter at hand. It was as if the moment for the perfect *elevator pitch*[91] had vanished. In that moment, I recognized that my ego was obstructing this new, unknown topic, and I inwardly laughed at its fears. I have suppressed my ego[92], the cough subsided, and I delivered the pitch flawlessly.

---

[91] *Presentation for the elevator* (or *speech for elevator*) (Eng. *Elevator Pitch* or *Elevator Speech*) is a short story about the concept of a product, project or service. The term reflects a limited time — the length of the presentation should be such that it can be fully told during the elevator ride, that is, one to two minutes.

The term is usually used in the context of the entrepreneur's presentation of the concept of a new business to a partner of a venture fund to obtain investments. Since representatives of venture capital funds strive to make a decision as soon as possible about the prospects or futility of a particular project or team, the primary selection criterion is the quality of the "presentation for the elevator". Accordingly, the quality of this speech and the level of its presentation are of paramount importance for the head of a startup seeking to find funding.

A well-designed elevator presentation answers the following questions:
- what product is offered,
- what are the advantages of the product,
- Information about the company.

[92] If you follow the Kabbalistic theory, your ego will begin to recede after 62–64 years, and before that, your victories over the ego are temporary and within a maximum of three months, the ego restores itself in its rights after your next victory over it. I happened to see my ego from the outside and even go into it from the almost no visible ego to its current level during deep meditation, and I realized for

Internal resistance is not the enemy; it is an overprotective friend who aims to assist but sometimes does the opposite. Yet, if handled skillfully, this integral part of human psychology can be advantageous. Inner resistance is tied to your ego — a blessing and a curse. Reflect on how absurd it sounds to proclaim, "I used to do this, but I managed to conquer my ego." Your ego grows with you and persistently resists your development — don't be fooled.

Ponder this thought separately from the rest.

---

myself the following: 14 billion years of our development allowed us to take only a few children's steps and, in general, the situation with our awareness is still simply terrifying and, as I saw it for myself, we have developed no more than two percent of what is possible.

Chapter VIII

Confronting
External Enemies

"You have enemies? Good. That means you've
stood up for something, sometime in your life."
Winston Churchill[93]

External enemies, who can they be? A Kabbalistic tale recounts a father taking a sharp knife from his son, which the son uses for crafting shoes. The father explains that in due time, he will happily introduce his son to the craft and entrust him with all the necessary tools, including the knife. However, for safety reasons, he chooses to take the knife away for now. The son feels upset, not fully comprehending his father's narrative. Only many years later does the true meaning of the incident become clear. At times, the Universe takes dangerous tools from our hands, and we hastily accuse it of neglecting our needs. Random individuals who defend their interests, as they perceive them, can also be classified as enemies.

For instance, in New York, parking is a nightmare. The city administration, playing populism for the elections, implemented markings that reduced parking spaces by 10%. Suddenly, an already acute situation became even worse. One summer evening, I parked near my house and had to wait a few minutes for a taxi to depart before I could park. With me in the car was a girl for whom I had certain plans. However, she seemed hesitant. When the taxi finally left, I signaled my turn and began to park. Just as I completed the maneuver and turned off the engine, a red car stopped near me, and the man inside started cursing, demanding that I free up a parking space for him since, as it turned out, he was also waiting for it.

The irony of the moment was that there was enough space for both our cars to park together. I immediately

---

[93] Sir Winston Churchill was an inspirational statesman, writer, orator and leader who led Britain to victory in the Second World War. He served as Conservative Prime Minister twice — from 1940 to 1945 (before being defeated in the 1945 general election by the Labour leader Clement Attlee) and from 1951 to 1955.

pointed this out. However, fueled by aggression, the man's ego prevented him from rational thinking, and he continued to angrily spew curses. Considering my companion's presence in the car, the situation grew ugly. I proposed that he step out of his car so that we could settle the matter like men. By that, I meant calmly and judiciously explaining the situation and, if necessary, resorting to force as a last resort in response to aggression. The man eagerly agreed.

As I stepped out of the car and waited, he popped open his trunk and retrieved a baseball bat. Equipped with the bat, this rather imposing figure charged toward me. Oddly enough, I found the situation hilariously comical from the attacker's perspective. The disproportionate aggression, accompanied by a weapon, threatened the man with years in prison regardless of the outcome. Moreover, in close-quarters combat, the bat posed more problems than solutions. It was evident that he was bound to fail unless I yielded to the threat. Either way, his defeat was imminent if I chose the path of maximum physical danger for myself or, as described further in this book, if I chose death.

In all honesty, I burst into laughter. I remained "in the armor of my shrines"[94], committing no wrongdoing at any level, and suggested peacefully resolving the issue. In essence, all the trump cards were in my hands. Regardless, the man found himself on the losing end, realizing that I had no intention of backing down without a fight. A fight in which he had already lost due to his unwarranted aggression. The man in the red car was completely caught off guard and stared at me in astonishment. While I continued to chuckle, I assumed the open stance of hachi

---

[94] ... In the lineage of the hippopotamus am I found,
    Clad in the sacred armor of my faith's embrace,
    With solemn strides, my spirit unbound,
    Through vast deserts, fear I shall not face.
    Nikolay Gumilev.

ji dachi — facing the attacker, arms slightly spread to the sides. The aggressor tossed the bat aside and fled, even leaving his car in the middle of the road. Once he drove off, I picked up the discarded weapon from the pavement and assisted the girl out of the car. Needless to say, she no longer had any doubts about spending the evening with me, and I acquired a new piece of sports equipment. A month later, in a similar fashion, I became the proud owner of three beautiful ceramic knives after encountering a gang of youths in Brooklyn.

This entire anecdote is not solely, and not primarily, about the external enemy — it goes deeper. It delves into the realm of the internal enemy. Don't allow your inner enemy — the ego — to conspire with the external enemy. By doing so, you will face a pure situation with a straightforward solution.

However, there are enemies who actively seek to take away your life, property, and harm you or your loved ones. They are not random individuals subconsciously seeking a noradrenaline[95] rush by acting against you. More often than not, they are a group of people acting consciously and purposefully, following a well-prepared plan, skillfully attempting to frame their assassination attempts as accidents, inexplicable acts of aggression, or crimes with motives other than their true intentions.

They are the ones who strive to hinder your thoughts, ideas, themes, words, and actions through targeted violence. Usually, you pose no immediate threat to

---

[95] *Noradrenaline* is a hormone that is involved in the implementation of fight-or-flight reactions, but to a lesser extent than adrenaline. The level of norepinephrine in the blood rises under stressful conditions, shock, injuries, blood loss, burns, anxiety, fear, nervous tension. This hormone may be responsible for the general activation of brain activity (for example, inhibition of sleep centers), an increase in motor activity, a decrease in pain sensitivity (plays the role of an anesthetic), an improvement in learning ("teaches" to defeat dangers), positive emotions ("a sense of victory").

them, but your worldview threatens their future as they perceive it. When you act in alignment with your beliefs, you are on the right path — there is no greater validation of your actions than having enemies who wish to take your life for them.

I am grateful to have such enemies in my life. Nevertheless, I am not prepared to surrender my life to them without a fight. Discipline and cautiousness prove helpful in battling these enemies. Experience shows that avoiding outright carelessness often allows one to identify immediate danger. The main principle, therefore, is to assess the threat before it materializes and take minimal precautions by applying the principle of "do not assist."

The topic has been extensively studied, but in short, it is advisable to adhere to the principle of "*red zones*" — avoiding places where practical security cannot be ensured and favoring controlled spaces instead, where enhanced security measures are in place. Pay attention to unusual or inexplicable circumstances, especially those that restrict your movements, prevent a quick escape, or eliminate the possibility of communication interruption in the face of unexplained behavior.

Do not assist your enemies, no matter how cliché it may sound. Do not alter your intentions to appease someone else's interests or even your own ego. The U.S. Constitution defines treason[96] as waging war against America or aiding and supporting enemies. How many times have we witnessed a country that has been attacked continuing to trade with the aggressor country? In general, history is filled with such moments.

---

[96] Article III, Section 3, Clause 1: Treason against the United States, shall consist only in levying War against them, or in adhering to their Enemies, giving them Aid and Comfort. / Article III, Section3, Paragraph 1: Treason against the United States will consist only in unleashing war against them or in joining their enemies in aiding and abetting them.

A logical question arises: what should one do when attacked by an enemy, facing a direct and immediate threat to their life or the lives of their loved ones, and pacification measures prove ineffective or there is no time for them? The answer is simple and effective: eliminate the threat.

**Now, carefully consider everything,
leaving no decision
to the heat of the moment.**

Chapter IX
Divine Trials

T he concept of divine trials often intertwines with spiritual and religious beliefs, highlighting the presence of a higher power that oversees the universe and individual destinies. These trials signify the challenges and obstacles we must overcome to foster personal growth and development.

Throughout history, numerous examples of divine trials have emerged as methods to determine guilt or innocence, as well as to settle disputes. In the Middle Ages and earlier cultures, trials often relied on chance or Divine Providence to ascertain the rightness or degree of guilt. Despite many instances of abuse, the pursuit of truth remains evident.

One such trial, known as *Trial by Battle*, involved physical combat between two individuals to determine the veracity of their claims. The belief held that God would intervene, ensuring victory for the innocent side. Another practice, *the Water Test*, subjected the accused to immersion in water. If they drowned, they were deemed innocent; if they surfaced, guilty. This method was commonly used in medieval Europe to identify witches.

Trial by Fire was yet another ordeal, where the accused had to walk through fire or hold a red-hot iron. Unharmed, they were presumed innocent; burned, they were declared guilty. This practice, similar to the Water Test, aimed to identify witches in medieval Europe. Additionally, *the Poison Test* involved administering a poisonous substance to the accused. If they survived, they were considered innocent; if they died, guilty. Various cultures, including ancient India, utilized this method.

It is crucial to note that these methods of divine judgment often relied on superstition rather than concrete evidence, resulting in the wrongful accusation and punishment of many innocent individuals. Over time, more rational and fair approaches have replaced these practices. Despite the glaring injustice of the earlier methods, progress has been made.

In many spiritual traditions, these trials are viewed as necessary steps on the path of self-development and spiritual purification. They offer opportunities to learn vital lessons, develop new skills, and ultimately become better individuals.

The nature of these trials may vary depending on beliefs and cultural context. Some trials may involve physical activities such as fasting, enduring extreme temperatures, or engaging in intense physical training. Others may focus on emotional or psychological challenges, such as confronting fears, overcoming negative thought patterns, or practicing forgiveness and compassion.

Regardless of their nature, these experiences are often transformative, enabling us to break free from old habits and patterns and adopt new ways of living that align with our highest potential. By facing these challenges with courage, perseverance, and faith, we develop the inner strength and resilience needed to navigate the highs and lows of life.

Ultimately, these trials serve to purify our spirits and bodies, shedding what no longer serves us and preparing us for the next stage of our spiritual journey. This may involve attaining deeper levels of understanding, compassion, and wisdom, or simply learning to embrace the present moment more fully.

In essence, the concept of divine trials reminds us that life is a path of growth and evolution, where challenges and obstacles are integral parts of our journey. By embracing these challenges with an open mind and heart, we cultivate the qualities and virtues necessary for a meaningful and purposeful existence.

The practical aspect of divine trials also holds relevance in the search for happiness. How do we endure these trials internally? In Japanese tradition, there exists a concept called "*gaman*"[97]. Gaman embodies the ability to endure the unbearable with patience and dignity. Rooted in Zen Buddhism, it reflects perseverance, resilience, and stoicism in the face of adversity. Accepting one's destiny forms a fundamental aspect of Japanese culture.

Throughout Japanese history, gaman has helped individuals navigate numerous challenges, including natural disasters, wars, and economic hardships. The experiences of Japanese Americans during World War II exemplify gaman in action. Despite the unjust internment camps they were subjected to following the attack on Pearl Harbor, they endured with patience and dignity. They faced harsh living conditions, loss of homes and possessions, constant surveillance, and discrimination. Through their perseverance, they preserved their cultural identity and dignity.

The aftermath of the 2011 earthquake and tsunami in Japan provides another testament to gaman. Despite widespread destruction, loss of lives, displacement, and the destruction of homes, businesses, and infrastructure, the Japanese faced the immense trials with endurance, patience, and dignity.

The resilience and determination of the Japanese allowed them to rebuild their communities and lives, demonstrating the power of gaman in the face of adversity.

In conclusion, gaman is deeply ingrained in Japanese society as a cultural concept, embodying perseverance, resilience, and stoicism when confronting adversity. It has proven to be a powerful force, guiding the Japanese people

---

[97] *Gaman* (我慢) is a Japanese term derived from Zen Buddhism, meaning "to endure the unbearable with patience and dignity." As such, it is a kind of stoic stamina, although it also requires maintaining self-control and discipline during difficult times in order to weather the storm in the best possible way. The haman is not weakness or submissiveness, but rather a demonstration of strength in the face of the adversity and suffering that it usually entails.

through numerous challenges in history and inspiring them to this day.

Based on personal experiences, I have encountered trials that I wouldn't willingly endure again. In some instances, I embraced gaman, while in others, I neglected it. Reflecting on those situations, I found that practicing gaman improved my self-perception, expedited resolutions, and minimized losses. Perhaps the calmness and clarity that gaman provides facilitate rational decision-making and deter potential adversaries. Other factors may also contribute.

Nicholas Roerich[98], a philosopher, traveler, writer, and artist, once proclaimed, "Blessed are the obstacles — we grow by them." He also advocated for a Peace Pact, suggesting cultural property be marked with a special flag of peace — a red ring with three red dots. The intention behind this proposal was to safeguard cultural values, schools, and museums from destruction during future conflicts. In essence, the idea transcends immediate contexts of conflict, manifesting as a creator who recognizes the creator within others and prioritizes future generations over self-interest. A prosperous society arises when individuals invest in the future, even if they won't reap the rewards of their labor.

Let us contemplate if this mirrors the likeness of the Creator.

---

[98] *Nikola Constantinović Rerich (Röhrich)* (September 27 [October 9] 1874, St. Petersburg – December 13, 1947, Naggar, Himachal Pradesh, India) was a Russian artist, set designer, mystic philosopher, writer, traveler, archaeologist, public figure. Academician of the Imperial Academy of Arts (1909). During his lifetime, he created about 7000 paintings, many of which are in famous galleries around the world, and about thirty volumes of literary works, including one poetic. The author of the idea and initiator of the Roerich Pact, the founder of the international cultural movements "Peace through Culture" and "Banner of Peace".

Chapter X

Confronting Peripetias

The term "peripeteia" derives from the ancient Greek word περιπέτεια, meaning "sudden change" or "conversion." Its root lies in the Greek verb (περιπέτεια), signifying a sudden reversal of fortune, an abrupt shift from good luck to misfortune.

In ancient Greek literature, twists and turns were commonly employed as a technique in tragedies, where the protagonist would experience an unexpected change of fate, often due to their own actions. This turning point would ultimately lead to their downfall or a tragic end, eliciting fear, pity, and awe in the audience.

The concept of twists and turns has influenced the development of narrative and drama throughout history. Aristotle's[99] Poetics[100] defines peripeteia as one of the essential elements of tragedy, alongside plot, character, thought, diction, and spectacle. The notion of a sudden twist of fate has found its place in various forms of storytelling, ranging from ancient Greek tragedies to modern films and novels.

Within the realm of vicissitudes, the Hellenic tradition highlights the significance of Hybris (Θύμβρις) — the personification of audacity and pride. According to Greek legend, the god of war, Polemos, relentlessly pursued Hybris, leading to the belief that she did not erect sanctuaries in cities or settlements. It was believed that

---

[99] *Aristotle* (384–322 BC) was a Greek philosopher and polymath of the classical period in ancient Greece, a vivid example of the "universal man". Trained by Plato, he was the founder of the Peripatetic school of philosophy at Lycea and the broader Aristotelian tradition. His writings cover many subjects, including physics, biology, zoology, metaphysics, logic, ethics, aesthetics, poetry, theater, music, rhetoric, psychology, linguistics, economics, politics, meteorology, geology, and public administration.

[100] *Poetýka* (from the Greek ποιητική; implied τέχνη *"poetic lawsuit"*) — the theory of poetry. The science that studies poetic activity, its origin, forms and meaning, and more broadly, the laws of literature in general.

after people displayed impudence, war would inevitably ensue, as the Greeks explained.

It was widely acknowledged that a hero's hybris would inevitably result in vicissitudes, as the gods withdrew their favor from the audacious, causing an upheaval in their fortunes.

In contemporary usage, the term "twists and turns" often describes any sudden or unexpected change or shift in events, transcending the boundaries of literature and drama. It encompasses situations where there is an abrupt alteration in circumstances, significantly impacting the final outcome.

In real life, twists and turns can also refer to unforeseen changes or events that profoundly affect an individual's life or circumstances. In practical terms, it is wise to temper one's pride, considering our limited knowledge, as exemplified by the Kabbalistic tradition's formula of 1% knowledge and 99% ignorance. With such an understanding, exercising caution in matters of pride seems only reasonable.

However, when confronted with inevitable vicissitudes, strategies to weather the storm come into play:

Stability

Resilience enables individuals to adapt and recover from adversity. Those who possess resilience are better equipped to navigate twists and turns by acknowledging the situation, accepting their emotions, and taking proactive steps to move forward. Inner peace becomes the foundation of resilience, attainable through reconciling emotionally with the worst possible outcomes while cherishing any positive developments as miracles.

Flexibility

Embracing new ideas and different approaches helps individuals adapt to changing circumstances. Those who exhibit flexibility are more likely to discover solutions when faced with twists and turns.

Planning

While planning for every contingency proves impossible, having a plan in place can mitigate the effects of twists and turns. Individuals with contingency plans, emergency funds, or backup strategies are better prepared for unforeseen events. On a personal level, this may involve saving 10% of earnings and maintaining a dedicated account for contingencies, as discussed in this book.

Support Network

Building a strong support system makes a significant difference when confronting twists and turns. Whether it comprises friends, family, or a professional network, relying on dependable individuals helps manage emotions and find solutions.

Positive Thinking

Maintaining a positive mindset enables individuals to focus on the opportunities that arise amidst twists and turns. Rather than fixating on the negatives, those with a positive outlook are more inclined to seek solutions and discover new avenues for personal growth.

Engaging in discussions with wise individuals further expands our understanding of these strategies. The concept of the universe as a continuous struggle between energy and nothingness, with the expansion of the universe ensured through a chain of conscious ideas or Logos, resonates deeply. This chain operates on different vibrations, starting with a conscious idea (Logos) and descending to lower vibrations, encompassing emotions, the energy of light (translated into photons), quarks, gluons, and the recently discovered Higgs boson.

The Higgs boson, also known as the Higgs particle, is an elementary particle in the Standard Model of particle physics. It arises from the quantum excitation of the Higgs field, one of the fields in particle physics theory. As a massive scalar boson with zero spin, even (positive) parity, no electric charge, and no color charge, the Higgs particle interacts with mass but is highly unstable, decaying into other particles almost instantly upon generation.

It is through these elementary particles that matter, in the form of protons, electrons, and neutrons, comes into existence, each with a lower vibration than the preceding element. This intricate chain of vibrations and energy transformations plays a vital role in the creation and sustenance of the universe.

Given the precious nature of emotions as energy units for future stars, it becomes imperative to establish a conscious mechanism that protects against the destruction or wastage of these resources. This mechanism safeguards the universe's development and prohibits actions that would hinder its progress.

By implementing such a mechanism, the universe gains the ability to foresee what may be detrimental to its development and actively prevents such occurrences. This foresight becomes a guiding force, ensuring the efficient utilization of resources, including emotions, and directing them towards their highest purpose.

In the cosmic struggle between energy and nothingness, the interplay of conscious intention and the protection of resources becomes a delicate dance. Those who initiate wars, for example, act directly against the development and expansion of the universe. The idea of war appears foolish and counterproductive, as it contradicts the universe's pursuit of harmony, growth, and enlightenment.

Those who commence destructive conflicts sow the seeds of their own peripeteia — a sudden change in fortune brought about by their misguided actions. The universe, in its grand design, seeks unity, interconnectedness, and the preservation of balance. It responds to disruptive actions with consequences, working to restore equilibrium and redirect the course of events.

In light of this understanding, embracing peace, compassion, and cooperation aligns with the purpose of the universe. By embodying these values, we contribute to the unfolding story of the universe and its relentless pursuit of expansion and enlightenment.

Allow yourself to remember this!

Chapter XI

Avoiding of Peripetia

The Art of Navigating Life's Twists and Turns

L ife, with its capricious nature, thrusts upon us the unexpected — those sudden shifts of fate that alter our course without warning. But fear not, for amidst this chaos, there exist means to mitigate the impact of these vicissitudes and reduce their frequency.

Arrogance Control

One pivotal aspect to ponder is the control of arrogance, that pinnacle of hubris which often leads to our downfall. Excessive pride, intertwined with our fear of not obtaining what we desire, manifests itself in the rudeness we exhibit towards others. In The Grand Budapest Hotel, M. Gustave astutely remarked, "Rudeness is merely the expression of fear. People fear they won't get what they want. The most dreadful and unattractive person only needs to be loved, and they will open up like a flower." These profound words shed light on the intricacies of human behavior, exposing the ego's craving for superiority and the desire to assert dominance over our fellow beings.

Yet, upon reflection, I have come to realize that this approach lacks efficacy. In my own experience, yelling at someone has never persuaded them of my righteousness or altered their convictions. The act of expressing anger and resorting to rudeness requires little courage. True audacity lies in love — a courageous choice to trust in the goodwill of others and embrace our shared humanity. To opt for love over hostility demands far greater *chutzpah*[101].

---

[101] *Chutzpah* (/ˈxʊtspə, ˈhʊt-/) is the quality of audacity, for good or for bad. It derives from the Hebrew word ḥuṣpāh (חֻצְפָּה), meaning "insolence", "cheek" or "audacity". Thus the original Yiddish word has a strongly negative connotation but the form which entered Englishas a Yiddishism in American English has taken on a broader meaning, having been popularized through vernacular use in film, literature, and television. The word is sometimes interpreted — particularly in

Upon deeper introspection, I recall moments of great rudeness in my own life. Fear, I now understand, was at the core of those regrettable outbursts. Fear of exclusion, of being forgotten or harmed. It is during these moments when our fear is most pronounced that we gravitate towards aggression, mistakenly believing it will shield us from harm. Yet, there exists a nobler path, a higher response.

Allow me to recount a profound incident that left an indelible impression upon me. Picture a busy street in the City, the stage for an unfolding brawl. One man viciously assaulting a taxi driver on the famous West End Avenue in New York, driven by an irrelevant and petty dispute. The consequences for both parties involved were bound to be severe, with their lives teetering on the brink of destruction. Is this not a reflection of what humanity does on a larger scale, time and again?

Regrettably, our society suffers from a dearth of moral courage. We bear witness to the ferocity of animal courage, while the fortitude to act with love and compassion remains scarce. Anger and pride become the enemies of non-violence, devouring our noble intentions. Only the brave are capable of displaying true love, for cowardice and profound emotion cannot coexist.

To fight arrogance which is a continuous process of course, one can pursue various avenues, such as immersing oneself in the accomplishments of science, art, and literature. Through these channels, one can witness the intellect, strength, abilities, courage, and perseverance of others. Additionally, travel presents a formidable method. By embarking on journeys, time seems to decelerate, and each new experience obtained serves as a step toward broadening one's knowledge, expanding the self. The allure lies in the inevitable encounters with evidence of remarkable achievements made by others, starting from

business parlance—as meaning the amount of courage, mettle or ardor that an individual has.

the mode of transportation and continuing as you explore different sights. It is essential to consciously acknowledge the positive aspects of fellow human beings, thus achieving a harmonious equilibrium.

Pride, an overabundance of confidence in one's abilities or position, finds its place in the spotlight as both a driving force for villains and main characters. Over the course of thousands of years, people have observed the punishment inflicted upon the proud. This archetype of punishment serves as a deterrent against the squandering of energy and resources. Pride leads individuals to make illogical decisions, acting against their own interests. Such behavior hampers the expansion of the Universe, which thrives on the development fostered by positive emotions[102]. These emotions serve as the fuel for the creation of stellar energy, the formation of matter, and ultimately, the perpetuation of life itself. In a sense, a positive inner attitude acts as an internal quantum shield against negative incidents. This aspect holds significance, and it deserves further exploration as, essentially, we are literally made of the ancient emotions through the chain of the transformation of energy to the matter[103]. The idea that positive thinking

---

[102] I will intentionally not delve into the context of the importance of positive thinking in this instance, but my beliefs are based on fragmentary knowledge from more than 10000 different books that I have read in my life, my life and my near-death experience paint me something like this: the basis of the universe is thought, which is transformed into the energy of light when the vibration decreases, with a further decrease in vibration, this energy is transformed into elementary particles, then into atoms and, sequentially, into matter. This process has its own laws, which are associated with the effects of acquiring speed, lowering speed, changing the frequency of vibration and gravity, however, the relationship between thought and matter is obvious to me.

[103] Like *Quarks*, for example. When bombarding subatomic particles, their presence is detected by modern instruments. Assuming that quarks are a lower vibration transformation of the Universe energy

can have a quantum impact on our lives has gained popularity in recent years. However, it's important to note that classical science, at the time of writing this book, does not fully endorse this claim. Therefore, let's exercise caution in our judgments. Nonetheless, experiments conducted at the renowned Large Hadron Collider[104] have, to this point, confirmed this assertion: separated particles respond when they encounter the particles they were once connected to. Quantum mechanics, a branch of physics that delves into the behavior of tiny particles such as atoms and subatomic entities, introduces a key concept known as the observer effect. It suggests that the mere act of observation can influence the behavior of a system. However, it's worth noting that this effect has only been reliably documented at the quantum level. When considering more complex aspects of the universe, ethical challenges arise, and the observer effect does not extend to larger macroscopic objects and systems. Nevertheless, it is intriguing to consider that since ancient times, humans have observed the phenomenon of separated twins who,

---

created from the gravitation force of the matter in the everexpanding Universe, the link between the emotions and the matter should be obvious.

[104] *Large Hadron Collider*, abbreviated *LHC* (eng. *Large Hadron Collider, LHC*) is a charged particle accelerator on colliding beams, designed to accelerate protons and heavy ions (lead ions) and study the products of their collisions. The collider was built at CERN (European Council for Nuclear Research), located near Geneva, on the border of Switzerland and France. The LHC is the largest experimental facility in the world. More than 10,000 scientists and engineers from more than 100 countries have participated in construction and research. "Large" is named because of its size: the length of the main ring of the accelerator is 26,659 m; "hadron" — due to the fact that it accelerates hadrons: protons and heavy nuclei of atoms; "Collider" (eng. collider — collider) — due to the fact that two beams of accelerated particles collide in opposite directions at special collision points — inside elementary particle detectors.

despite being separated at birth, often display strikingly similar behavior, tastes, and, at times, even fate[105].

Motivated by their desire to elevate themselves, the proud seek to raise their social status and shape others' opinions to align with their own self-perception. Persuasion, showboating, and relentless practice become the means through which they achieve their goals. However, the proud often succumb to delusions of grandeur. Merely believing oneself to be exceptional at tennis does not equate to actual skill, and this misguided belief can lead them astray.

The flaw inherent in proud characters lies in their self-centeredness. They become the focal point of their own universe, oblivious to the existence and contributions of those around them. This myopia hinders their ability to recognize the value others bring to the table, resulting in the exclusion of teamwork from their plans. Moreover, their hubris blinds them to their own shortcomings, propelling them headlong into failure and downfall.

In the profound words of Gandhi, "In a gentle way, you can shake the world." Let us choose the path of gentleness and compassion, for it is through these virtues that we find the strength to confront life's twists and turns. As we navigate the eternal struggle between energy and nothingness, let us acknowledge the conscious mechanism that safeguards our emotions and resources. May we develop foresight to discern what impedes the universe's growth and expansion. In doing so, we become contributors to the enlightenment and flourishing of the cosmos itself.

---

[105] Watch the documentary "Three Identical Strangers," which tells the story of three twin brothers who were separated in infancy and given to different families. David, Robert and Edward met quite by chance as adults and for a long time tried to find out why they were separated. It turned out that the brothers became part of a psychological experiment that was conducted in the 60s on a dozen twins.

Those who believe themselves invincible and above the rules often succumb to unnecessary risks that ultimately lead to their downfall. Acknowledging our limitations and embracing humility become the keys to inspiration. We can find enlightenment by studying the achievements of others, immersing ourselves in the realms of science, art, and the written word. Through travel, time slows, and we bear witness to the remarkable feats of humanity, humbling reminders of our collective brilliance.

Acceptance of uncertainty

Embracing uncertainty stands as another vital tool in navigating life's twists and turns. Uncertainty is intrinsic to life itself, and unexpected changes become our steadfast companions. By embracing this uncertainty, we cultivate flexibility, openness to new ideas, and the resilience necessary to adapt when the ground beneath us shifts.

Building resilience

Building resilience emerges as an essential component in our journey. A positive attitude, bolstered by a network of supportive friends and effective coping mechanisms, enables us to rebound from setbacks and triumph over challenges. Within ourselves lies a quantum mechanism of protection, a positive mindset that serves as a shield against negative occurrences. While the scientific validation for the quantum effects of positive thinking remains inconclusive, experiments in quantum mechanics have demonstrated the interconnectedness of once-unified particles. The phenomenon of separated twins further underscores the existence of unseen forces at play.

While we cannot meticulously plan for every eventuality, preparing for contingencies can help mitigate the impact of sudden shifts. Establishing a financial safety net, formulating contingency plans for potential job loss or major life changes, and securing insurance coverage serve as proactive measures. Such planning brings us closer to reality, grounding us and deterring reckless actions born of arrogance.

Seeking advice and feedback from trusted sources assumes paramount importance. Through diverse perspectives, we unveil blind spots, evade the pitfalls of overconfidence, and make informed decisions. Wise counsel becomes our compass, steering us away from treacherous waters.

In conclusion, while we cannot fully evade life's twists and turns, we possess the power to diminish their effects. By taming our arrogance, embracing uncertainty, building resilience, planning for contingencies, and seeking guidance, we chart a course that allows us to navigate the unpredictable seas of existence. Peripetia may still lurk, but armed with wisdom and humility, we can confront it head-on.

The choice lies within us — to succumb to fear and hostility or to defy them with love and courage. Let us become the architects of a harmonious existence, where resilience accompanies peripetia, and the flourishing of the universe is ensured through the power of our conscious ideas and the energy of our emotions.

That's what I wanted to tell you about it.

Chapter XII

Principles of Doing Business

Business — the act of entrepreneurship, carried out with personal or borrowed funds, taking risks and shouldering responsibility, with the primary goals of making a profit, or money[106] and developing one's own enterprise. All other activities we pursue under the guise of business, without the aim of generating profit, are merely hobbies[107]. The crux of the matter lies in profitability and the accumulation of wealth. Certain traditions have cast doubt upon business and those engaged in it.

However, in the Judaic tradition, business is not regarded as a necessary evil, but rather as a divine calling. It posits that business activity possesses inherent sanctity and can serve as a catalyst for personal transformation and positive impact on the environment and all those involved.

A notable discussion rooted in Kabbalah poses a question after one's passing: "Did you conduct business in good faith?" This inquiry implies that there exists a moral good to be found in the manner in which business is conducted. It underscores the significance of engaging

---

[106] I was surprised how many people fail to understand exactly what is money, so *money* — is a universal equivalent, acting as a measure of the value of goods or services, easily exchanged for them (having maximum liquidity). In its form, money can be a special commodity, a security, a sign of value, various goods or valuables, account entries. The main functions of money are distinguished: a measure of value, a means of circulation, a means of payment and a means of accumulation.

[107] *Hobby* (from the English. *hobby* — hobby, favorite thing) or hobby — a type of human activity that is engaged in leisure, for enjoyment. Passion is something that a person loves and is happy to do in his free time. Infatuation is a good way to deal with stress, depression, anger, anger or rage. In addition, hobbies often help to develop horizons. The main purpose of hobbies is to help self-actualization. Hobbies are divided into 3 main types: making things, collecting things, and learning things.

in business with integrity and creating genuine and meaningful value.

This same tradition draws examples from the Jewish heritage, such as the actions of Jacob in establishing stable currency, local markets, and infrastructure developments. These examples illustrate how seemingly ordinary business activities can be elevated to moral acts. It suggests that engaging in business can be a path towards righteousness and self-sufficiency, a means to create a narrative of transformation through the pursuit of good faith business endeavors. It compels individuals to contemplate how they can craft the most compelling story of their daily business interactions and how their enterprises can effect positive change in the world. Ultimately, it exhorts individuals to conduct business with integrity, generate value, and contribute positively to society.

Other traditions also imply similar sets of moral principles, urging individuals to carefully consider and embrace the valuable lessons that resonate with them.

There exists, however, an aspect often overlooked by economists in relation to money, which, in my view, holds paramount importance in understanding its true nature. I have read books that vilify money, and I have read books that extol its virtues. Yet, money, like a katana[108], possesses no inherent moral quality. Its capacity for good or evil lies solely in the hands of its wielder. It is my profound conviction that money acts as a vessel for the energy of its possessor. It can be used wisely, for personal benefit and the betterment of society, in which case it multiplies. Conversely, it can be squandered foolishly, serving only the whims of one's ego, resulting in its departure. I have dedicated several years to studying the most prominent American families who lost their

---

[108] *Katána* (Jap. 刀) is a long Japanese sword.

fortunes during and after the *country's Golden Age*[109], as well as those who managed to expand their wealth.

In all cases, a consistent pattern emerged: those who centered their use of money around personal indulgence, constructing unimaginably luxurious palaces, traveling extravagantly, and acquiring lavish goods, ultimately experienced loss. On the other hand, those who approached their acquisitions with the mindset of actively and honestly engaging in philanthropy, building schools, establishing public parks, and aiding in nature conservation, emerged as winners. There exists an ancient Japanese principle that partially elucidates this phenomenon.

Ikigai[110], a Japanese concept denoting the search for purpose or reason for being, offers a valuable perspective

---

[109] *Golden* or *The Gilded Age* (Eng. *Gilded Age*) is an era of rapid growth of the economy and population of the United States after the Civil War and the reconstruction of the South. The name comes from the book "The Gilded Age" by Mark Twain and Charles Warner and plays on the term golden age, which in American history was gilded only on the surface. It is believed that the modern American economy was created in the era of the "Gilded Age". In the 1870s and 1880s, both the economy as a whole and wages, wealth, national product, and capital in the United States grew at the fastest pace in the country's history. Thus, between 1865 and 1898 wheat crops increased by 256%, corn by 222%, coal production by 800%, and the total length of railway tracks by 567%. The corporation has become the dominant form of business organization. By the beginning of the twentieth century, per capita income and industrial output in the United States had become the highest in the world. Per capita income in the United States was twice as high as that of Germany and France, and 50% higher than that of Britain. In the era of the technological revolution, businessmen built new industrial cities in the Northeast of the United States with city-forming factories and plants, which employed hired workers from different European countries. Multimillionaires such as John D. Rockefeller, Andrew Mellon, Andrew Carnegie, John Morgan, Cornelius Vanderbilt, the Astor family gained a reputation as robber barons.

[110] *Ikigai* (Jap. 生き甲斐 ikigai, "meaning of life") is a Japanese concept meaning a sense of one's own purpose in life; It can be a hobby, a

when contemplating the principles of conducting business. From four distinct viewpoints, it sheds light on the essence of the concept:

Employee

For an employee, business principles encompass job security, fair compensation, and opportunities for growth and advancement. Through an ikigai lens, an employee can find fulfillment by aligning their work with personal values and strengths. This may involve seeking a position that allows them to utilize their skills and talents or finding meaning in the mission and purpose of the company they serve.

Businessman

From a businessman's stand-point, business principles encompass profitability, growth, and market share. However, an ikigai perspective urges businessmen to also strive for creating value for customers, employees, and society as a whole. By providing products or services that fulfill genuine needs or solve problems, a businessman can discover purpose and satisfaction in their work.

Sole Proprietorship

The principles of a self-employed individual may include autonomy, flexibility, and the pursuit of personal passions. Through an ikigai lens, a self-employed person seeks to align their work with their values and interests, whether by establishing a business that reflects their passions or by utilizing their skills and talents in a manner that positively impacts the world.

---

profession or a family. Ikigai is a philosophy that promotes longevity, helping to find satisfaction, joy and awareness in all things every day. Thanks to ikigai, a person learns to put his thoughts in order, notice the beauty of the world around him and enjoy the little things, finds harmony and peace of mind.

Investor

An investor's perspective on business principles may encompass risk management, diversification, and profit maximization. However, an ikigai perspective encourages investors to support companies that align with their personal values and beliefs. By investing in enterprises that contribute to the betterment of the world, investors can find purpose and fulfillment in their financial pursuits.

Traditionally, the approach to employment involved seeking stable jobs that offered security and long term benefits. However, this paradigm has shifted in recent times, with employees now seeking opportunities for self-realization and growth.

Google stands as a prime example of success in this regard. It consistently ranks among the best places to work, thanks to its employee-centric policies and culture. Google provides various benefits and perks, such as free meals, onsite fitness centers, and flexible working hours, fostering the overall well-being of its employees. Additionally, Google's innovative and collaborative work environment empowers each team member to leverage their skills and strengths to the fullest.

Conversely, the retail giant Walmart serves as an example of failure in prioritizing employee well-being. It often faces criticism for offering low wages, subpar working conditions, and limited opportunities for growth and advancement. Many employees report feelings of being overworked and underpaid, resulting in high turnover and diminished morale.

For businessmen, success is typically measured in terms of profitability, growth, and market share. However, businesses that prioritize these goals at the expense of ethics and social responsibility may eventually meet failure.

Patagonia serves as a compelling example of success in this regard. The outerwear company has managed to achieve both profitability and social responsibility. Patagonia demonstrates a firm commitment to environmental protection by donating 1% of its sales to environmental causes. Moreover, the company maintains a transparent and ethical supply chain, fostering trust and loyalty among its customers.

In contrast, Enron Corporation stands as an example of failure. Once one of the largest energy companies globally, Enron's unethical and illegal accounting practices led to its eventual collapse in 2001. Several top executives faced bankruptcy and conviction as a result.

For self-employed individuals, success often stems from the ability to pursue their passions and create value for others. However, financial instability and burnout pose constant threats.

J. K. Rowling, the author of the Harry Potter series, embodies success in this regard. Initially faced with multiple rejections from publishers, Rowling persevered and eventually achieved remarkable success. She has become one of the most influential writers of all time, inspiring an entire generation of readers and filmmakers.

On the other hand, Elizabeth Holmes, the American entrepreneur and founder of Theranos, exemplifies failure. Holmes claimed to have developed ground-breaking blood testing technology capable of detecting various diseases with just a single drop of blood. However, it was later revealed that these claims were false, leading to the company's bankruptcy.

For investors, success hinges upon effectively managing risk and maximizing returns. However, investors may also seek to align their investments with their personal values and beliefs.

AP7, a Swedish pension fund, serves as a notable example of success in this regard. The fund has divested from fossil fuels and instead invested in renewable energy. AP7 consistently outperforms its competitors in terms of return on investment, while also fostering the development of innovative technologies.

Honesty holds significant importance in business, as it cultivates trust and long term relationships with customers, suppliers, and stakeholders. However, honesty must be balanced with the principle of trade secrets.

The concept of trade secrets acknowledges that businesses possess the right to safeguard certain information to protect their intellectual property, competitive advantages, and trade secrets. This principle proves particularly relevant in industries like technology, pharmaceuticals, and manufacturing, where innovation and intellectual property play crucial roles.

While honesty remains vital, businesses must also find a balance between transparency and accountability, and the protection of commercial interests. This often leads to ethical dilemmas and tensions, especially concerning data privacy, product security, and supply chain management.

For instance, a company may discover a potential defect in a product that could harm consumers. While there exists an ethical obligation to disclose this information and take corrective action, the company may hesitate due to concerns about reputation or the potential exposure of trade secrets to competitors. In such cases, businesses must navigate the delicate balance between honesty, transparency, and protecting their commercial interests.

Non-disclosure agreements[111] (NDA's) further complicate the balance between honesty and trade secrets. While these legal agreements are common practice across various industries, they can also be misused to conceal unethical or illegal behavior, such as fraud or discrimination. In such instances, businesses must weigh

---

[111] *Non-Disclosure Agreement.* A basic agreement that protects trade secrets. There is a type of NDA — NDCA (*Non-Disclosure and Non-Circumvention Agreement*) — a document that prevents access directly to the asset in question.

the need for confidentiality against the importance of honesty and accountability.

Furthermore, the interplay between honesty and trade secrets can be influenced by external factors like regulations and public opinion. The technology industry, for instance, has witnessed an increasing demand for transparency and accountability regarding data privacy and algorithmic use[112]. This has led to heightened control, regulation, and public pressure on technology companies, compelling them to be more open and honest about their activities.

> Additionally, several core elements underpin successful business organization. Whether working for a company or planning to establish one, understanding these eight fundamental business principles is crucial for achieving success.

1. *Sufficient Quality Product*: A business must offer a product or service of top-notch quality that instills pride. Just as Emile Berliner's gramophone represented cutting-edge technology in its time, your business needs a product that stands out in the market.

2. *Know Your Industry and Competitors*: Even the finest product can fail if there is no demand or if it lacks competitive pricing. Emile Berliner faced rivalry from Edison, whose legal maneuvering drove Berliner's telephone invention out of the market. Understanding the industry and the

---

[112] For example, Elon Musk opened the patents of his company Tesla and invites everyone to use the information protected by patent law if it will be used for the fair use of Tesla technology, thus helping his competitors due to the importance of switching to the use of electrical energy for transportation.

His own words on this subject sound like this: "*Technology leadership is not defined by patents ... but rather by the ability of a company to attract and motivate the world*".

players within, including those prone to playing dirty tricks, is essential to assess how your offering will fare against the competition.

3. *Promote Your Products and Services*: Thoroughly study the probability of completing the full range of services for your clients. Familiarize yourself with the fundamentals of marketing through traditional and digital media to effectively share your company's products and services. Invest ample time in preparing a comprehensive marketing plan as a worthwhile investment for growing your business.

4. *Build a Great Staff*: Your employees constitute your most valuable asset. They can either help build a successful company or contribute to its downfall. Hire and train your staff so that they embody your mission, goals, and care for them almost as much as you do.

5. *Understand Organizational Structure and Design*: A basic understanding of how well-functioning businesses operate is essential. From establishing departments and personnel to managing individual projects, aim to run your business effectively and efficiently.

6. *Adequate Capital for Business Development*: Business is not a mere hobby; it must generate profit. While starting with a solid financial base is crucial, maintaining positive cash flow as you grow and develop is equally important. Prepare a comprehensive business plan, regardless of your current investors' eagerness to provide sufficient funds.

7. *Fundamental Principles of Accounting and Finance*: Comprehend and adhere to the fundamental principles of accounting and finance, along with the applicable regulations and laws within your

industry. Obtain necessary permits, fulfill tax obligations, and diligently observe filing deadlines.

8. *Respect Your Customers*: While customers may not always be right, it is essential to go above and beyond to rectify any issues and provide them with the best possible products and services. Treat them with respect and understanding, striving to maintain their satisfaction.

Take some time to reflect on these thoughts.

Chapter XIII

Principles of Relationships with Loved Ones

Our immediate family members, especially our mother and father, have a significant impact on shaping our life's trajectory. In many instances, we can observe patterns of behavior that guide us towards making conscious decisions about our own lives, essentially teaching us valuable life lessons.[113] Childhood experiences serve as the foundation for the development of our future adult selves. The emotional connection with our mother is vital, as a lack thereof can potentially lead to difficulties in establishing and maintaining our own family relationships. On the other hand, the absence of a strong emotional bond with our father may hinder our ability to take calculated risks, venture into business endeavors, and achieve financial success.

These harmful patterns within the family dynamic can pose a threat to our ultimate goal in life: the pursuit of happiness. They can impede our journey towards self sufficiency and hinder our ability to provide for the needs of our loved ones. It is important to examine the potential consequences of providing too much for our loved ones as well. While the theme of family dynamics and their impact is vast, we will focus specifically on the perils that may hinder our potential for happiness.

By exploring these dynamics and patterns, we can gain a deeper understanding of the challenges we may face in our pursuit of happiness. Recognizing and addressing these patterns can help us break free from their negative influence, allowing us to forge our own path towards fulfillment and contentment.

Most of our behaviors are formed during childhood, where our parents play a vital role in shaping our

---

[113] I do not claim that my reading of the Kabbalistic texts is correct, but I believe that there is a certain internal logic in the choice of those souls who unite on earth in families — the level of inner affection and unconditional love, it seems to me, indicates the presence of fundamental internal contradictions that such souls must overcome on their way.

interactions with others, understanding our desires, and evaluating the world around us. The dynamics within the family have a profound impact on a child's worldview, including how they perceive themselves.

Option 1: Disregarding Everyone

In this scenario, parents are disconnected from each other and their child. The father comes home from work, and his family doesn't engage with him. The mother is absorbed in social networks, while the children watch cartoons. This emotional detachment leaves the child feeling disconnected from the world. Without witnessing their parents' communication and problem-solving, they lack a framework for interacting with others. These children may find solace in connecting with nature and animals.

Option 2: Father and Child

This type of relationship involves the mother being isolated, and the child primarily interacting with the father. The father distances himself from the mother and openly disrespects her. This behavior is equally harmful to children, regardless of their gender. Girls may subconsciously suppress their femininity to avoid becoming targets like their mother, while boys may emulate their father's attitudes and actions.

Option 3: Mother and Child

In this scenario, the father is isolated, and the child's interaction is mainly with the mother, who is detached from the father. The girl in such a family may perceive herself as an extension of her mother, inhibiting her ability to embrace her femininity. Since the importance of men is consistently downplayed, a grown daughter may struggle to develop a fulfilling intimate relationship, as her mother subconsciously restricts her from embracing "adult sexuality." The boy, on the other hand, may become emotionally dependent on his mother and struggle to mature.

Option 4: Mother and Father, but not Child

Parents communicate with each other but are emotionally distant from the child. This often occurs when young parents entrust the upbringing of their child to grandparents. The lack of parental involvement from the father leads the child to subconsciously seek a mentor or teacher. The absence of maternal nurturing can result in difficulties in identifying the female component in children and foster negative attitudes towards their own bodies. The child's fate can take two paths: they may later choose a significantly older partner or completely abstain from love relationships, or they may mature quickly, feeling as though their childhood was stolen, and seek solace through games, computers, or gambling.

Option 5: We're Only Together Because of You

Parents in this situation are in conflict and contemplate divorce but remain together for the sake of the child. Each parent maintains communication with the child. Subconsciously, the child feels guilty, reinforcing beliefs like "I can't make anyone happy" or "I'm to blame for everything." They may also experience deep-seated suppressed anger towards their parents. However, individuals from this type of relationship often develop strong negotiation skills, becoming diplomats, teachers, and the like.

Option 6: Dad is the Head of Everything

In this dynamic, the father becomes the central figure in the relationship, interacting with both the child and the mother. However, there is no intimacy between the mother and child. This idealizes the father, leading to a subconscious female confrontation between the mother and daughter for the main man's attention. The father subconsciously discourages his son from surpassing him, promoting rivalry and hostility. As the boy grows up, he may exhibit the same aggressive attitude towards other men.

Option 7: Our Mother is the Best in the World

Here, the mother becomes the connecting link in the family, communicating with the child and the father. However, the child's access to the father is solely through the mother. In such a relationship, a girl may unconsciously restrict her communication with men, while a boy may develop animosity towards his father, perceiving himself as the only worthy partner for his mother[114].

While these relationship types rarely occur in their pure forms, they significantly impact individuals in adulthood. The key to addressing this situation lies in awareness. It is essential to understand the suppressed emotions and feelings towards our parents, verbalize them, and provide an outlet for expression. While we cannot change our parents or other people, we always have the power to change our attitude towards them. Recognizing that they loved us and granting them the right to make mistakes can help build a new, harmonious image of our family within ourselves. This transformation allows for constructive relationships with the outside world.

---

[114] *Oedipus complex*, in psychoanalytic theory, a desire for sexual involvement with the parent of the opposite sex and a concomitant sense of rivalry with the parent of the same sex; a crucial stage in the normal developmental process. Sigmund Freud introduced the concept in his Interpretation of Dreams (1899). The term derives from the Theban hero Oedipus of Greek legend, who unknowingly slew his father and married his mother; its female analogue, the Electra complex, is named for another mythological figure, who helped slay her mother.

Parental alienation[115], a phenomenon where a child is manipulated to reject a parent, often arises during divorce or custody battles but can occur in intact families as well. It involves one parent conveying exaggerated or false information to the child, leading to the child refusing a relationship with the other parent. This experience can be deeply distressing for the child, causing confusion, sadness, and a sense of loneliness. They may struggle to understand their conflicting emotions and develop self-blame due to the manipulation. Parental alienation can severely restrict the child's time with the targeted parent and involve negative comments, false accusations, and threats to withhold affection. This type of forced-estrangement refers to an induced undesired change in the child's perceptions and feelings about the target parent similar in process to the *Stockholm Syndrome*[116].

---

[115] When children are asked to carry hostile messages from one parent to another, are exposed to negative and belittling comments about the other parent, or are forbidden from mentioning the other parent, they are placed in loyalty conflicts which result in significant stress (Grych, 2005; Kelly & Emery, 2003). Cummings, Goeke–Morey, and Papp (2001) describe this kind of involvement in hostile disputes between parents as "destructive conflict." Research informs that one parent's denigration of the other parent or pressure exerted on the child to withdraw from the other parent can lead to confusion, self denigration, or a complete rejection of the other parent (Cummings, Goeke–Morey, & Papp, 2001; Grych, Harold, & Miles, 2003; Grych, 2005). Leona Kopetski (1998a, 1998b) refers to this as a form of psycho-social pathology exhibited by alienating parents who manage their internal conflicts through contrived interpersonal conflict based on exaggerated or nonexistent character flaws in the alienated parent. She learned from her research with over 600 families (1975–1995) that this psycho-social pathology invariably leads to a type of forced-estrangement between the child involved and the target parent who is unjustifiably demonized by the alienating parent (Kopetski, 1998a, 1998b).

[116] In 1973 in Stockholm Sweden the Kreditbanken Bank was robbed. The robbers held captive several bank employees for six days while

Parental alienation is not officially recognized as a diagnosable condition by the psychological community, but the act of brainwashing a child against a parent is acknowledged and can be addressed in court if robust evidence is presented. However, false allegations of alienation can also occur for custody or financial purposes.

Parental alienation can stem from a parent's reliance on their child for emotional support, intensifying during a divorce. The child, wanting to support the alienating parent, may come to believe and internalize false claims. Other motivations behind parental alienation include revenge, jealousy, and financial extortion. This behavior is often associated with narcissistic and emotionally unstable parents who transfer their pain and rage onto the child.

Preventing parental alienation after a divorce is best achieved when parents maintain amicable relationships and avoid denigrating each other. Seeking support outside of the child, having sensitive discussions away from their presence, and accepting new partners can contribute to the child's stability and happiness.

When an expartner turns a child against the other parent, the targeted parent can gather thorough evidence, such as witness testimonies and communication records, to address the issue in court. Legal processes may involve

---

attempts were made to negotiate with police. During this period, the bank employees became emotionally attached to their captors. In so doing, they refused assistance at one point from government sources, sang their captors praises, defended them and even refused to participate in their criminal prosecution, all of this after they were released from their ordeal. The bank employees, all adults and strangers to their captors prior to the robbery, developed a profound change in their thinking and their emotional experience of their captivity and their captors. This change moved from fear to affection. By terms of common parlance, they became brainwashed as a result of forced indoctrination while being held captive.

psychological evaluations, custody assessments, family assessments, and reunification therapy to rebuild the relationship.

Estimates suggest that parental alienation occurs in 11 to 15 percent of divorces involving children, with around 1 percent of children in North America experiencing it. Fractured relationships resulting from parental alienation can often be healed over time through reunification programs and therapy.

Parental alienation is not a criminal offense but rather a matter dealt with through civil proceedings. Some debate exists regarding whether it should be criminalized due to the lasting damage it causes, but proving parental alienation can be challenging and it is not yet considered a diagnosable syndrome.

While parental alienation is not listed as a disorder in the DSM-5, some suggest it falls under "parent–child relational problem" that merits clinical attention. It is seen by some researchers as a form of emotional child abuse and family violence.

Societal and legal changes that could help prevent parental alienation include fostering amicable co-parenting relationships, raising awareness about the damaging effects of alienation, and providing resources for mental health support to families going through divorce or separation.

Healing from parental alienation can be a long and challenging journey. Ostracized parents should know they are not alone and strive to express compassion and kindness toward the estranged child. Seeking support from friends, family, support groups, or mental health professionals can be beneficial during this process. Gradually increasing the child's time with the targeted parent and refraining from denigrating the alienating parent can help repair the relationship. Individual therapy for all parties involved can aid in healing the trauma caused by parental alienation.

Children who experience parental alienation may struggle with self esteem, guilt, self hatred, depression, and substance use. Repairing the relationship with the child often involves limiting their time with the alienating parent and increasing time with the targeted parent. It takes time for the child's biased view to clear, but research shows that even severely damaged relationships can be repaired.

As adults, individuals may come to recognize that they were victims of parental alienation. The process can be emotionally painful and may take years or even decades. Understanding the signs and strategies of parental alienation, speaking with the targeted parent, and seeking professional guidance can help in identifying if one was a victim of alienation.

Repairing a broken relationship with a parent as an adult can be challenging. However, gaining a better understanding of the parent's perspective and the broader situation can help foster reconciliation. Communication and empathy are essential, and it is important not to denigrate the alienating parent or dismiss the child's feelings during this process.

In some cases, estranged adult children may consistently exhibit hostility or threaten legal action, indicating that it may be time for the targeted parent to stop attempting to reconnect. Each situation is unique, and it is essential to prioritize one's own well-being and mental health.

The notion of bread of shame originates from the Kabbalistic tradition, portraying an energy imbalance that manifests itself initially as embarrassment and discomfort when receiving something unearned. When it comes to our interactions with family members, the concept of bread of shame can serve as a guiding principle, urging us to embrace respect, gratitude, and reciprocity. Neglecting this principle will inevitably lead to contempt from our relatives, who will gladly allow us to toil for them without a shred of gratitude[117], eventually culminating in hatred[118]. Strange as it may sound, this outcome is both inevitable and, in fact, justified from their perspective. Personally, I have experienced the effects of bread of shame, yet I cannot claim to have fully learned this lesson; I have merely become aware of the extent of my own ego[119].

> Here are a few principles that can aid in communicating with relatives within the context of bread of shame:

Respect

Demonstrate respect for your relatives by acknowledging their inherent worth and value as individuals. Refrain from

---

[117] Following the same Kabbalistic tradition, any benefits that you provide to any people, including relatives, without a commensurate, I emphasize, commensurate remuneration, which can be achieved in one or more of the following three ways: 1) A sincere depth of gratitude corresponding to your action; and/or 2) Counter service; and/or 3) Counter commensurate material benefits will lead to the provision of bread of shame and hatred for you. Moreover, this hatred will be absolutely justified towards you, since you have deprived them of the opportunity to make the necessary overcoming themselves.

[118] To clarify: in the Kabbalistic tradition, hatred is a desire for someone's physical death. I emphasize, not necessarily planning, but at least a strong desire.

[119] In the Kabbalistic tradition, our work on ourselves is completely endless — we only tear off another veil of the ego, for which the analogy of onion petals is used. For each new layer of the bulb there will be the next layer. To clarify: my meditations show me that there are at least 200 billion years of the development of the universe ahead, taking into account the fact that for almost 14 billion years we have passed about 2% of the way.

belittling or humiliating them, even during moments of disagreement or conflict. This test of wisdom will require your utmost strength. Remember that everyone possesses their own strengths and weaknesses, and strive to appreciate your loved ones for who they are. It will demand all your effort to support them without robbing them of their desire to resemble the Creator. Bear in mind that any inclination to provide everything for them without their own efforts is merely the game of your ego, intent on your destruction and that of your loved ones.

Gratitude

Express gratitude towards your relatives for the support they have offered you, be it emotional, financial, or otherwise. Show your appreciation through words or gestures, and never take their contributions for granted. Do not overlook the context of bread of shame in relation to their assistance to you; hasten[120] to demonstrate gratitude and reciprocate the favors bestowed upon you.

Reciprocity

Strive to reciprocate the kindness and generosity of your relatives by offering your own support and assistance in return. This may entail aiding them with tasks or projects, lending a listening ear when they require someone to confide in, or simply spending quality time together.

Boundaries

Maintain healthy boundaries with your relatives by establishing clear limits regarding what you are willing to do or tolerate. Respect their boundaries in turn and refrain from overstepping boundaries that may cause discomfort or breed resentment.

---

[120] The Kabbalistic tradition speaks of three months that are given to us in order to avoid the bread of shame when receiving this or that benefit.

Forgiveness

Practice forgiveness with your relatives by releasing past grievances and focusing on the present and future. This does not entail disregarding or condoning harmful behavior, but rather finding constructive and positive ways to move forward.

By adhering to these principles, we can cultivate relationships with our relatives grounded in mutual respect, gratitude, and reciprocity, rather than succumbing to the uncomfortable and perilous sense of shame on a physical level.

**I encourage you to approach my words critically
and engage in your own reflections
to evaluate their validity.**

Chapter XIV

Principles of Relations
with Strangers

Principles of interactions with unfamiliar individuals can be distilled into two main categories:

1. When engaging with strangers whom you neither assist nor conduct business with, it is crucial to adhere to specific principles for your own safety and well-being. Consider the following guidelines:

Respect personal boundaries

Regardless of familiarity, it is vital to honor others' boundaries. If someone chooses not to engage or accept your assistance, respect their decision without persistence. Each person possesses a right to privacy and personal space. Reflect upon the state of Texas, known for its intricate history within the United States and its liberal stance on firearm ownership, adopting "Friendship" as its motto.

Exercise caution

While interacting with strangers, exercise prudence and safeguard yourself from potential risks. Refrain from disclosing personal information such as your address or phone number, and be cautious when asked for money or other services.

Trust your instincts

If something feels amiss, trust your instincts and intuition. Did you suddenly spill water while conversing? Did you experience a stomachache or sudden nausea? Temporarily postpone decisions and take immediate measures to ensure your safety. If you feel uneasy or threatened, remove yourself from the situation or politely decline offers of assistance.

Practice politeness

Even if you cannot offer help or engage in business, maintain politeness and respect. Treat others as you would like to be treated and avoid rudeness or dismissive behavior.

Be clear and direct

When someone persistently seeks assistance or attempts to conduct business, communicate your boundaries with clarity and firmness. Politely but firmly inform them that you cannot provide the requested help.

> Overall, when interacting with strangers whom you do not assist or engage in business with, prioritize your safety and well-being. Treat others with respect and kindness, placing yourself first.

MEN'S SAFETY

Let us now explore the concept of exercising reasonable caution for adult males:

Several principles govern safe conduct for adult men to protect themselves, their loved ones, and their possessions. These principles encompass:

Awareness

Always remain cognizant of your surroundings. Stay vigilant, alert to potential threats such as suspicious individuals or activities in your vicinity.

Preparedness

Be ready for potential threats by having contingency plans in place. This could involve designating a safe room or shelter in your home that your loved ones are aware of, establishing communication plans with family members or neighbors, and maintaining an emergency kit containing essential supplies like food, water, first aid materials, and backup communication tools.

Self-defense

Acquire self-defense skills, such as martial arts, or carry authorized self-defense equipment to protect yourself and your loved ones during attacks. Numerous everyday objects can serve as weapons when necessary. For instance, tightly gripping a cluster of keys can enhance the force of a strike, while folding a matchbox in half and holding the resulting sharp edges between your fingers creates a potent blow to the throat, stomach, eye, or groin, delivering a painful shock. When facing multiple assailants, identify the leader and strike ruthlessly. Prepare yourself for the possibility of sustaining injuries during an attack, as this mental readiness aids in overcoming the initial shock.

Risk assessment

Evaluate the level of risk in different situations and take appropriate precautions. For example, when driving through unfamiliar areas, research the local crime rates and adopt additional measures to protect yourself. Remember, in unfamiliar places, it is wise to divulge as little personal information as possible. Draw the curtains in motels and exercise caution.

Avoidance

Steer clear of dangerous situations or individuals to ensure your safety. This may involve avoiding specific locations at night, refraining from interactions with suspicious individuals, or abstaining from risky behaviors such as excessive alcohol or drug consumption in public places.

Communication

Maintain open and transparent communication with family members or neighbors to prevent dangerous situations. Ensure everyone is familiar with the contingency plan and knows how to contact each other in emergencies. Commit the phone numbers of your loved ones to memory rather than relying solely on phone contacts.

Technology

Leverage technology for security purposes. Install a home security system, utilize GPS trackers on your phone, or employ personal safety apps as tools to safeguard yourself and your loved ones.

By adhering to these principles of safe conduct, adult males can proactively protect themselves, their loved ones, and their property. Remember, security is an ongoing process necessitating constant vigilance and attention to effectively anticipate potential threats.

WOMEN'S SAFETY

Several principles guide safe conduct for adult women to protect themselves, their children, and their possessions. These principles encompass:

Awareness

Always remain cognizant of your surroundings. Stay vigilant, alert to potential threats such as suspicious individuals or activities in your vicinity.

Preparedness

Be ready for potential threats by having contingency plans in place. This may involve having a dedicated safe room or shelter in your home, establishing communication plans with family members or neighbors, and maintaining an emergency kit containing essential supplies like food, water, and first aid materials.

Self-defense

Acquire self-defense skills, such as martial arts, or carry self-defense equipment like pepper spray or a stun gun to protect yourself and your loved ones in case of an attack. Remember that escaping from attackers is the ideal outcome.

Risk assessment

Evaluate the level of risk in different situations and take appropriate precautions. For example, when traveling to unfamiliar areas, research the local crime rates and adopt additional measures to ensure your safety and that of your children.

Avoidance

Steer clear of dangerous situations or individuals to ensure your safety. This may involve avoiding certain places at night, refraining from interactions with suspicious individuals, or abstaining from risky behaviors such as excessive alcohol or drug use. Be cautious of leaving food or drinks unattended in bars or nighttime venues. Control the attention you attract and maintain a safe distance. It is acceptable to err on the side of caution when interpreting strangers' intentions.

Communication

Maintain open and clear communication with family members or neighbors to prevent dangerous situations. Ensure everyone is familiar with the contingency plan and knows how to contact each other in emergencies.

Technology

Leverage technology for security purposes. Install a home security system, use a GPS tracker on your phone, or utilize personal safety apps as tools to protect yourself and your loved ones.

Trust your intuition

Women often possess a strong intuition or gut instinct in dangerous situations. It is crucial to trust and act upon these feelings, taking steps to safeguard yourself and your children.

> By adhering to these principles of safe conduct, adult women can proactively protect themselves, their children, and their property. Remember, security is an ongoing process necessitating constant vigilance and attention to effectively anticipate potential threats.

CHILD SAFETY

Teaching children about safe behavior is an integral part of parenting. It equips them with necessary life skills and prepares them to navigate various situations they may encounter in adulthood. Here are principles of safe behavior parents can impart to their children:

Stranger danger

Educate children about the potential risks associated with interacting with strangers or accepting gifts from them. Explain that they should never accompany strangers or enter a vehicle with an unfamiliar person.

Respecting personal boundaries

Teach children to respect their own personal boundaries and avoid unwanted physical contact. Emphasize their right to decline hugs, kisses, or any form of physical contact that makes them uncomfortable.

Road safety

Educate children about the significance of road safety, including how to cross streets safely, utilize crosswalks, and wear reflective clothing or accessories when walking or cycling during nighttime hours.

Water safety

Educate children about the dangers of water, be it swimming pools, lakes, or oceans. Emphasize the importance of adult supervision while swimming and the necessity of wearing a life jacket while boating.

Cybersecurity

Educate children about the potential dangers of the internet and social media, such as online predators, cyberbullying, and identity theft. Emphasize the importance of keeping personal information confidential and reporting any suspicious activity to an adult.

Emergency preparedness

Teach children about the significance of being prepared for emergencies, including what to do in case of fires, earthquakes, or other emergencies. Ensure they know how to contact emergency services and establish a designated meeting area for emergencies.

First aid

Educate children about basic first aid skills, such as treating minor injuries like cuts, bruises, and burns. Ensure they know how to seek help in the event of more serious injuries or emergencies.

Personal responsibility

Instill in children the importance of taking responsibility for their own safety, including avoiding risky behaviors, using safety equipment like bicycle helmets, and speaking up if they feel unsafe or uncomfortable.

> By teaching children these principles of safe behavior, parents can empower them to make wise and safe choices throughout their lives, enabling them to handle various situations effectively.
>
> To evaluate the effectiveness of these principles, one could ask individuals who have experienced violence if having known and practiced these principles beforehand could have helped them. These principles serve as valuable tools to navigate the digital world and outsmart adversaries lurking in the shadows.

Here are also ten personal cyber security tips:

1. *Keep your software up to date.* Patch the old, vulnerable software, and keep your devices secure. Automatic updates are key. Remember, even the most stubborn folks can change their ways.
2. *Use antivirus protection and a firewall.* Protect your data from the malicious invaders. Build a wall around your digital fortress, and let no enemy pass.

3. *Create strong passwords and manage them wisely.* Forget the complex mix of characters. Choose something user-friendly yet secure. Never repeat passwords or leave hints for the enemy.

4. *Embrace two-factor or multi-factor authentication.* Go beyond the standard password. Add layers of security to keep the hackers at bay. Don't rely on SMS; they can exploit it.

5. *Beware of phishing scams.* Be suspicious of strange emails, calls, or flyers. The enemy disguises itself, seeking to deceive. Don't be fooled by their tricks.

6. *Safeguard your sensitive information.* Don't reveal too much. Social media is a dangerous realm. Keep your personal identity hidden. Privacy is your shield.

7. *Secure your mobile devices.* Lock them tight. Set a difficult passcode, install apps from trusted sources, and keep your device up to date. Don't let the enemy infiltrate your pocket.

8. *Back up your data regularly.* Follow the 3–2–1 rule. Keep copies in different places, including the cloud. When ransomware strikes, erase and restore from your safe haven.

9. *Avoid public Wi-Fi without a VPN.* Encrypt your connection, and keep the cybercriminals guessing. If security is paramount, rely on your own network.

10. *Monitor your online accounts and credit reports.* Guard your financial life. Check for any suspicious changes. Lock your credit and hold the key to your kingdom.

Remember, knowledge and vigilance are your greatest weapons against cyber threats. Educate yourself and stay aware. Protect what is rightfully yours, and outsmart the enemy lurking in the digital shadows.

2.

When dealing with strangers whom you are providing assistance to or conducting business with for the first time due to the current circumstances, it is vital to adhere to certain principles to ensure your safety and well-being. Here are a few principles to bear in mind:

When helping strangers, it's crucial to acknowledge that you cannot offer assistance effectively if you neglect your own well-being. This principle revolves around the notion that you must prioritize taking care of yourself before being able to assist others.

> To illustrate this principle, consider the following examples:

Example 1: *A social worker disregarding their mental health.* A social worker engaged in aiding vulnerable populations may overlook their own mental well-being due to the emotional strain involved in their work. They may believe that helping others takes precedence over self-care. However, over time, this neglect can lead to burnout and emotional exhaustion, hampering their ability to provide effective support to their clients.

Example 2: *An overloaded therapist.* A therapist may feel compelled to take on an excessive number of clients to assist as many individuals as possible. However, if they fail to prioritize their own well-being and neglect to take necessary breaks, they may become overwhelmed and unable to deliver quality care to their clients.

Example 3: *A self-sacrificing volunteer.* A passionate volunteer committed to a cause may disregard their own physical needs, such as staying hydrated or taking breaks, in order to help to the maximum extent possible. However, if they become dehydrated or exhausted, they will be unable to continue providing assistance and may even require medical attention themselves.

In each of these instances, the principle of self-care takes precedence. Neglecting your own well-being can hinder your ability to effectively help others. It's important to recognize your own limitations and prioritize self-care to ensure you can continue providing assistance in the long run.

Nevertheless, there exists an astonishing and divine exception to this rule: extending aid in situations of extreme and urgent need.

While it is generally important to prioritize your own well-being in order to effectively assist others, there may be exceptions when you feel an intense and genuine sense of compassion for someone in dire circumstances. In such cases, you may set aside your own interests to offer immediate help and support to the person in need. This act can be viewed as a manifestation of selfless love or self-sacrifice.

Indications of extreme and urgent need for assistance include:

Life-threatening situations

When someone's life is in jeopardy, immediate action is required to save them.

Severe physical or mental disorders

If an individual is experiencing a severe physical or mental disorder, they may require immediate medical attention or psychological support.

Complete helplessness

If someone is unable to help themselves due to physical or mental limitations, they may require assistance with basic tasks such as eating, dressing, or bathing.

Extreme poverty or homelessness

Individuals living in extreme poverty or without shelter may require immediate support to meet their fundamental needs, such as food, shelter, and clothing.

In such exceptional circumstances, it may be necessary to temporarily set aside your own well-being and provide as much help as possible to the person in need. However, it's important to understand that sacrificing your well-being should not be a long term solution, as it can lead to burnout and exhaustion.

Allow me to recount a story involving a case that frequently appears in the biographies of wealthy individuals. One of the wealthiest people on Earth, personally known to me, hails from a small town located far from the centers of civilization. When this man sat for a crucial entrance exam at the university, where he aspired to be accepted but did not hold high hopes, he encountered a young woman with a child in an extremely perilous and challenging situation. To assist her, he thoughtlessly depleted his meager funds ear-marked for his journey. In order to afford the cheapest ticket to the exam location, he was compelled to work unloading railway cars for a day, with the money earned barely covering the cost. Exhausted from fatigue, he slept through the entire journey to the university instead of revising the material for the exam. However, during the exam, everything fell into a logical and comprehensible pattern for him. He successfully solved a highly complex problem in an original manner and was subsequently accepted.

Nonetheless, it is essential to exercise caution and carefully evaluate the situation before extending help. Make sure you are not putting yourself or others in harm's way by offering assistance. Seeking advice and guidance from trained professionals or authorities can help you take the most appropriate steps to assist the person in need while ensuring your own safety and well-being. Other principles of care will be discussed later in this book.

When interacting with someone for the first time in a situation where contact is necessary due to circumstances, such as making a purchase or conducting a sale, dining at a restaurant, or placing a food order, there are several principles you can follow to ensure a smooth interaction:

Be courteous and respectful

Regardless of your role as a customer, approach the interaction with a polite and respectful demeanor. This can foster trust and establish a positive relationship with the other person.

Communicate clearly and concisely

When engaging with the other person, employ clear and concise language. Avoid using jargon or technical terms that may be unfamiliar to them.

Listen actively

Pay close attention to what the other person is saying and ask clarifying questions if needed. This ensures that both parties have a clear understanding of the transaction at hand.

Negotiate honestly

When discussing prices or terms, offer fair and reasonable proposals. Consider the other person's needs and interests alongside your own.

Be mindful of cultural differences

If you are interacting with someone from a different culture, be aware of cultural disparities and adapt your communication style accordingly.

Deliver on your commitments

If you make an agreement or commitment with the other person, ensure that you follow through on it. This builds trust and fosters positive relationships for future interactions.

Overall, the key to interacting with someone for the first time in a buying or selling situation is to approach the encounter with a positive attitude, communicate clearly, and be willing to negotiate honestly. By adhering to these principles, you can establish a positive rapport with the other person and facilitate successful interactions.

Reflect on moments in your life when you did not adhere to these principles. With the passage of time, would you make the same choices again?

Chapter XV
Principles of Relations with Business Partners

The principles guiding relationships with business partners hold significant importance for the success of any company. In the vast expanse of this subject, I aim to outline the fundamental principles that have become the bedrock of the Age of Aquarius trends. Nurturing positive relationships with partners amidst these unprecedented circumstances can foster increased trust, enhanced communication, and a collaborative approach to problem-solving, which will be vital in the years to come. Conversely, negative relationships can lead to conflicts, missed opportunities, and tarnish a company's reputation. Here, I present the principles of relationships with business partners, accompanied by instances of positive and negative practices.

Principle No. 1
COMMUNICATION AND TRANSPARENCY

Effective communication and transparency play a crucial role in forging robust relationships with business partners. Clear and timely communication helps avert misunderstandings, identify issues early on, and ensures both parties are aligned towards common objectives.

Positive example:

The company diligently updates its investors on the progress of ongoing projects and shares pertinent information regarding its financial outcomes. Moreover, the company actively encourages investor partners to contribute their feedback and ideas, which are earnestly considered and integrated into future plans.

Negative example:

The company exhibits sluggishness in responding to inquiries from its business partners and investors, leading to frustration and bewilderment. Furthermore, the company withholds information that could impact partner activities, such as alterations in product lines or pricing strategies.

Principle No. 2
RESPECT AND TRUST

Respect and trust serve as crucial components in any fruitful relationship, including those between business partners. By valuing each other's opinions, experiences, and contributions, trust can be nurtured, fostering cooperation. To uphold trust and respect requires consistent effort, firmly rejecting manipulation even in trivial matters.

Allow me to recount a personal experience. I was hired to restructure a development business in New York focused on constructing modern affordable housing. Its previous collapse stemmed from the previous owners' fear of informing investors about the true course of events. As the business encountered difficulties, the management clung to untenable agreements until the situation reached a critical and irreparable point. Trust had been irrevocably shattered, and the business succumbed. Undoubtedly, there were reasons behind such management behavior, potentially linked to unwise investor conduct. Nonetheless, the company possessed the power to sever toxic relations in favor of the company's creditors and other investors. However, the company chose to postpone this course of action, ultimately resulting in its demise.

Positive example:

The company values the contributions of its business partners and actively seeks their input in crucial decisions. Additionally, the company upholds its commitments by meeting deadlines and delivering orders punctually.

Negative example:

The company frequently dismisses suggestions from its business partners or appropriates their ideas. Furthermore, the company fails to fulfill promises, such as providing adequate support or timely payments.

Principle No. 3
FAIRNESS AND EQUALITY

Business relationships should be built upon principles of fairness and equality. Both parties should feel that they derive mutual benefits from the partnership, with these benefits being distributed equitably.

Positive example:

The company offers fair prices and favorable conditions to its business partners for its products or services. Additionally, the company extends resources and support to assist its partners in succeeding.

Negative example:

The company takes advantage of its business partners by imposing unfair terms, such as unilateral contracts or exorbitant fees. Furthermore, the company prioritizes its own interests over those of its partners, fostering a sense of inequality.

Principle No. 4
FLEXIBILITY AND ADAPTABILITY

Business relationships undergo evolution over time, necessitating flexibility and adaptability from both parties to navigate market or industry changes.

Positive example:

The company collaborates closely with its business partners to identify novel opportunities and adjusts its strategies accordingly. Moreover, the company displays willingness to modify processes or products based on partner feedback.

Negative example:

The company resists change, refusing to consider new ideas or approaches. Additionally, the company disregards feedback from its partners, resulting in missed prospects and lost revenue.

Therefore, establishing positive relationships with business partners demands effective communication, respect, fairness, flexibility, and adaptability. Neglecting these principles can lead to detrimental consequences and tarnish a company's reputation. By prioritizing these principles, companies can foster strong partnerships that benefit both parties and contribute to long term success.

At a broader level, many corporations prioritize profits over the public interest, which can give rise to various adverse consequences for society. Some problems associated with this profitoriented mindset include:

Environmental degradation

Corporations that prioritize profits over the environment often engage in actions that harm our surroundings. Such actions can result in detrimental effects such as air and water pollution, deforestation, and climate change.

Exploitation of workers

Corporations that prioritize profits over the welfare of their workers may employ practices that exploit labor, such as low wages, inadequate working conditions, and insufficient benefits.

Harm to consumers

Corporations that prioritize profits over consumer well-being may partake in activities that harm consumers, such as selling harmful or addictive products.

Market monopolization

Corporations prioritizing profit above fair competition may employ methods that lead to market monopolization, which can be detrimental to consumers and small businesses.

Political influence

Corporations that prioritize profit utilize their economic power to influence policy decisions favoring their interests, often at the expense of the public interest.

These issues exemplify the negative consequences stemming from corporations prioritizing profits over the public interest. It is important to acknowledge that not all corporations adopt this approach, and there are numerous examples of companies that prioritize social and environmental responsibility over profits. Nevertheless, it is crucial to hold corporations accountable for their actions and encourage responsible and ethical business practices.

From my own experience, I have observed that the problem of unfair market exploitation and depletion of natural resources is deeply ingrained in society. Notably, leading universities demonstrate a cautious approach, ensuring they only maintain the necessary level of care to avoid direct prosecution while teaching their graduates how to manipulate markets. For instance, in prominent universities, the educational game "gas station" is prevalent. Although it offers valuable cognitive insights at various levels of perception, its primary message centers around market price manipulation without direct interaction with market participants, thus preventing regulators from proving corporate collusion in market manipulation. Ethically speaking, these principles are utterly destructive for the world's future[121].

**It is incumbent upon us all
to fully comprehend this reality.**

[121] This is perfectly seen in the prices of cellular services in the United States, where universities teach manipulation to their students and Germany, where it is less common: prices for the same service differ by 5 times, even though the quality of communication in the United States is many times worse than in Central Europe.

Chapter XVI

Principles of Action
in Conflict

In times of conflict, defending principles of justice and moral values is paramount, while establishing healthy boundaries with those involved becomes crucial. Sometimes, a point of no compromise emerges, rendering it impractical to continue. In such instances, it is wise to distance oneself from toxic individuals, relationships, or interactions.

Eleanor Roosevelt once remarked, "No one can diminish your worth without your permission." This notion highlights the power individuals possess to control their self-esteem and happiness by not allowing negative opinions to affect them. Stoic philosophy emphasizes the significance of mindset in overcoming adversity. We recognize the curriculum's connection to historical movements for equal rights, where individuals faced oppression but steadfastly pursued their goals. Thus, it is our duty and natural assignment to encourage educators to reintroduce these discussions to schools, starting from early childhood.

> To safeguard oneself and uphold values during conflicts, several actions can be taken, including:

Establishing boundaries

Clearly defining limits is essential in any relationship, even with toxic individuals. This entails communicating needs and expectations, identifying unacceptable behaviors, and outlining the consequences of crossing those boundaries. As my father taught me, giving someone three opportunities to correct their behavior, each time explaining alternative paths and their consequences, is sensible. However, enduring unfair treatment multiple times is illogical, with rules determined by the potential danger involved. For instance, life-threatening or health-threatening behavior should be promptly halted, irrespective of rules.

Prioritizing self-care

When dealing with toxic individuals, it is vital to prioritize one's own well-being. This may involve seeking support from friends or professionals, engaging in joyful or relaxing activities, and practicing self-compassion.

Seeking mediation or advice

If conflicts affect relationships with loved ones, seeking mediation or counseling can facilitate finding acceptable solutions for both parties. A neutral third party provides a safe space for communication, aiding in the identification of needs and the discovery of common ground.

Severing ties

In certain cases, it may be necessary to end a relationship with a toxic individual or group. While a challenging decision, it can be crucial for one's well-being and mental health. In my experience, the easiest path to making this decision is through internal reverence for one's divine purpose in this world.

Ultimately, the most appropriate course of action depends on the specific circumstances of the conflict. Prioritizing one's well-being and values while recognizing the needs and perspectives of the other person involved is paramount.

> Failing to negotiate, pursue legal action, seek protection from like-minded activists, or external support and thus failing to stand up for one's interests can have severe consequences for individuals and communities. Some common examples of self-destructive behavior in these areas include:

Inability to negotiate

When individuals or groups fail to reach an agreement, longstanding conflicts may arise, making it challenging to find common ground in the future and resulting in further polarization and ruptured relations.

Waiver of Judicial Protection

In certain situations, legal action becomes necessary to safeguard rights and interests. Failing to pursue legal recourse can lead to permanent violations of these rights, leaving individuals feeling helpless and frustrated. I have witnessed numerous cases where people's rights were systematically violated, despite their right to defend their interests in court. This often stems from people's inertia based on negative experiences with authorities, highlighting the shortcomings of the law enforcement system and the presence of corruption.

Failure to seek mediation and intervention from like-minded activists

Mediation and intervention by activists can be powerful instruments for promoting change and protecting rights. Neglecting to avail oneself of these resources can result in a lack of awareness and support for crucial issues, perpetuating harmful systems and practices.

Failure to seek external support

Seeking external support, such as counseling or therapy, plays a vital role in coping with stress, trauma, and other challenges. Failing to seek such support can lead to feelings of isolation, hopelessness, and exacerbate mental health issues.

Inability to take a position

When individuals or groups fail to stand up for their interests, they may face further marginalization and disenfranchisement. This can result in a loss of confidence, self-esteem, and purpose.

These examples of self-destructive behavior can have severe ramifications for individuals and communities.

Recognizing the value of negotiations, legal action, advocacy, activism, external support, and the promotion of positive change is crucial. By taking these steps, individuals contribute to personal and collective growth, fostering a more just and equitable society.

Handling difficult individuals can easily disrupt one's equilibrium. It could be anyone — a spouse, relative, friend, coworker, or boss. In this book, we delve into circumstances where people make unfavorable and disadvantageous decisions due to the distracting "cat on the scene" effect.

So, how should we navigate these troublesome individuals in our lives? It's tempting to respond with anger or defensiveness when they act maliciously or attempt to manipulate us, thinking it demonstrates strength. Yet, in reality, giving in to the impulse for retaliation reveals our weakness. Fighting fire with fire only fuels their toxicity, allowing their difficult behavior to persist.

A better approach lies in heeding the wisdom of the Jewish sage, *Ben Zoma*[122]: "Who is strong? The one who conquers their own desires." This means restraining ourselves from responding in kind and finding a way to react that respects both ourselves and the difficult person.

To move beyond annoyance and respond constructively, understanding the reasons behind their manipulative behavior is essential. Manipulation stands as a significant factor that robs one of happiness: numerous studies show that individuals with terminal illnesses often cite their inability to resist others' opinions and wishes regarding how their lives should be organized as the primary reason for dissatisfaction. Let's explore a few examples — guilt manipulators, issue collectors, and blamers — and how to address them:

---

[122] *Simeon ben Zoma*, also known as Simon ben Zoma, Shimon ben Zoma or simply Ben Zoma (Hebrew: בן זומא), was a tanna (kabbalistic teacher) of the 1st and 2nd centuries CE.

Guilt manipulators

Instead of directly expressing their needs and feelings, some people manipulate others into feeling guilty. They lack self-esteem, believing they don't deserve to ask for what they want. They play the role of martyr, hoping others will understand without requiring them to verbalize it. When their hopes are dashed, they become resentful or despondent. To deal with guilt manipulators, encourage them to express themselves clearly. Pose questions to clarify how they want their issues addressed.

Issue collectors

Certain individuals remain silent when upset, accumulating incidents until they unleash their frustrations all at once. They avoid making waves through timely complaints. Eventually, the pressure builds like a volcano, and they explode with pentup resentment. To handle such manipulators, acknowledge any discomfort you may have caused and apologize for your role. Let them know you prefer addressing issues as they arise instead of accumulating grievances.

Blamers

Some individuals engage in manipulative behavior rooted in their own fears of inadequacy. They accuse others of incompetence, using various aggressive and disruptive tactics to coerce recognition. Yielding to their manipulations closes the path to happiness for those who acquiesce. These individuals lack self-esteem and feel unsafe expressing themselves directly or taking responsibility for their actions. Insecure individuals who fear inferiority or abandonment resort to blaming or belittling others, hoping to erode their self-worth and discourage them from seeking better relationships. When faced with a blamer, collect yourself before responding. Stay calm and respectful, acknowledging any valid points they raise while explaining your perspective. If falsely accused, calmly state your disagreement and politely request specific examples to support their claims.

Compassion and common sense should guide us in dealing with difficult behaviors. If we can summon compassion for these challenging individuals, they'll sense it and feel less inclined to project their insecurities onto us. By employing positive communication techniques, we can become part of the solution.

However, it's crucial not to trap ourselves in toxic situations. If our efforts to diffuse the behavior prove futile, we must be willing to gracefully step away, at least temporarily. Remember, we cannot change someone else's behavior; we can only change our own. By exercising self-control when faced with difficult people, we demonstrate true strength of character. Responding with sensitivity fosters our well-being and theirs, recognizing that we are all in this together, facing our own challenges that provide opportunities for personal growth.

> Recall any toxic relationships in your life at various levels, and without getting distracted by their causes, reflect on how they developed and which consequences of your actions or inaction you consider most significant for your well-being.

Chapter XVII

Principles of Relations with Enemies

As expounded within these pages, the universe unveils itself as an eternal struggle between Everything and Nothingness, a perpetual contest for survival that reveals profound insights into the fabric of existence. At its core, this cosmic dance is driven by conscious ideas, forming a chain of Logos, each link pulsating with diminishing frequency.

Emotion, the zenith of this chain, emanates with a vibrancy distinct from the cerebral Logos, infusing the cosmos with its vital energy. It is the lifeblood that births forthcoming stars, as its lowered vibration begets the luminous particles known as photons — symbols of growth, transformation, and the perpetuity of cosmic expansion.

To safeguard and harness these invaluable reservoirs, the universe has erected a mechanism of consciousness, standing as a bulwark against the squandering and annihilation of emotions or, the fuel of the universe expansion. This mechanism acts as a vigilant guardian, guiding the universe's evolution and warding off deleterious outcomes.

In this framework, the necessity emerges for a prescient mechanism capable of foreseeing perils that threaten the progress of the cosmos. Armed with foresight, the universe can avert or discourage actions hindering its advancement. This extraordinary mechanism possesses a discerning eye, able to select actions, behaviors, or misalignments that impede universal development, ensuring that resources, including emotions, are directed towards their loftiest purpose.

Within the grand tapestry of cosmic struggle, the delicate interplay between conscious intent and resource preservation assumes a paramount role. By embracing this awareness and incorporating foresight, the universe acquires sagacity to navigate its course, shielding against the misuse or depletion of its precious energy.

Indeed, the folly and counterproductiveness of war become glaringly apparent. Those who ignite wars stand in direct opposition to the universe's growth and expansion. Through their engagement in such destructive conflicts, they sow the seeds of their own downfall, their own peripeteia — an abrupt change of fortune wrought by their own misguided actions.

The universe, resplendent in its magnificent design, aspires to harmony and flourishing, eschewing conflict and obliteration. It adheres to the principles of unity and interconnectedness, where every deed reverberates throughout the cosmic symphony. Those who disrupt this delicate equilibrium through war and aggression shall inevitably face the consequences of their choices. The universe, in its tireless pursuit of reinstating equilibrium and redirecting events, shall mete out justice.

Let it be known that the pursuit of war stands in stark contrast to the universal tapestry of development. It is a futile endeavor, yielding naught but suffering, devastation, and a profound detachment from the cosmic current. To align with the universe's purpose, we must embrace peace, compassion, and cooperation. In doing so, we contribute to the unfurling narrative of the cosmos, joining its relentless quest for expansion and enlightenment.

In light of this paradigm, for the sake of the universe's mere existence, and thus, our own, we must exert every effort to safeguard the wellspring of universal expansion — our positive emotions. It becomes apparent that engaging with those we perceive as enemies may deprive us of a wealth of positive emotions, hindering our development. Thus, the profound importance of handling our adversaries with care and discernment cannot be underestimated.

Relationships with enemies are often characterized by negative emotions such as anger, fear, and distrust. These emotions can be triggered by a perceived threat, real or imagined, and can make people withdrawn and irrational.

When we are in a state of negative emotions, our brain prioritizes the fight-or-flight response over reasoning and logical thinking. Blood recedes from the thinking part of the brain, the cerebral cortex, and rushes to the most ancient and primitive part of the brain, the "reptilian" trunk[123]. This can make it difficult to obtain new information and process it rationally.

In relations with the enemy, this can result in ignoring weighty arguments and focusing on victory at any cost.

---

[123] We have touched on this before, but it will not be superfluous to develop this topic slightly. The Triune Brain (Eng. Triune brain) is an evolutionary theory of the development of the human brain and mammals in general, rejected by science, in which three key components of the brain function relatively independently of each other: the brain stem, the limbic system and the neocortex — with which the name of the theory is associated. The theory offered an approach to understanding how the brain evolved in conjunction with its reactions under evolutionary pressure, and assumed a separate evolution of parts of the brain, in which new parts were layered on top of old ones in the course of evolution, as well as their independent functioning to some extent. From this position, at first, there was an evolution of behavioral reactions, then emotional reactions were added to them, then emotional reactions were supplemented by cognitive ones, including thinking, logic and planning. It should be noted that the author of the concept mistakenly assumed that new parts of the brain in the course of the evolution of vertebrates were added on top of existing, old areas, expanding the functionality of the brain. In fact, in all vertebrates, the brain can be divided into different-shaped forebrains, midbrains, and hindbrains, which traces evolution back to a common ancestor, meaning that major evolutionary changes concerned changing existing parts of the brain rather than adding new ones. The concept of the triune brain is one of the two main evolutionary theories of brain development, along with topological phylogenetics (parcelling theory), according to which new brain structures were not layered on top of old ones, but branched off from existing ones without disturbing the existing topology. However, regardless of the details of evolutionary development, the general assessment of the situation seems absolutely accurate to me, although not an expert, but having met with the practical consequences of the victory of the reptilian brain over the mind in specific circumstances.

Even if good ideas or arguments are presented, they can be rejected because of a negative emotional attitude towards the other person.

However, it is possible to change the dynamics of relations by directing aggression in a constructive direction. This may include finding common ground and working together to achieve a common goal that is beneficial to both parties. By adding rationality and reciprocity to the equation, acting pragmatically but coldly, it is possible to build trust and create more productive relationships, even if the underlying negative emotions still exist.

In general, the key to managing your relationship with your enemy is to recognize the impact of negative emotions and work to channel them in a constructive direction. By focusing on common goals and finding ways to build trust, it is possible to overcome the barriers created by negative emotions and create more positive relationships.

There are several examples of how enemies overcame their negative emotions and worked together to achieve a common goal, which led to positive results.

One example is the historic meeting between Mahatma Gandhi and Lord Irvine, Viceroy of India, in 1931. Gandhi was the leader of the Indian independence movement, and Lord Irvine was the representative of the British Empire. They met to discuss India's independence, were able to find common ground and work towards a peaceful settlement. They agreed to a compromise known as the Gandhi–Irwin Pact, which led to the release of political prisoners and paved the way for further negotiations for independence.

Another example is the collaboration between Apple and Microsoft in the late 1990s. At the time, the two companies were fierce competitors in the computer industry. However, they acknowledged that they could benefit from working together on certain projects. In 1997, Microsoft invested $150 million in Apple and committed to continuing to develop Office software for the Mac platform. This unexpected collaboration of competitors helped stabilize Apple's financial position and allowed Microsoft to expand its market share.

In the business world, there are many examples of former rivals joining forces to create successful partnerships. For example, in 2014, Lyft announced a partnership with the Chinese giant Didi Chuxing. At the time, Didi Chuxing was competing with Uber in the Chinese market, and Lyft was a smaller player in the U.S. market. However, by working together, they were able to create a network of ridesharing services spanning multiple continents.

In all these examples, the parties involved were able to overcome their negative emotions towards each other and find common ground. In this way, they were able to create partnerships and agreements that led to positive outcomes for all involved. These examples demonstrate the power of cooperation and the potential benefits of letting go of hostility and working toward common goals.

There are also many examples in history and business where negative emotions and lack of cooperation between enemies have led to negative results.

One historical example is the Cold War between the United States and the Soviet Union. During the Cold War, the two countries were in a state of intense hostility and rivalry caused by ideological differences and geopolitical concerns. This hostility led to a nuclear arms race and numerous proxy wars, including the Vietnam War and the Korean War. The Cold War ended with the collapse of the Soviet Union, but tensions and mistrust between the two countries had a lasting impact on global politics, which again resulted in open confrontation through the Russian Federation's hostile attack on Ukraine in 2014.

In 2014, the Russian Federation's aggressive actions towards Ukraine, accompanied by the rise of a fascist regime stemming from a kleptocracy that had its origins in the flourishing organized crime of the 1990s, have raised alarming concerns about the potential for global conflicts in 2022. The escalation of the conflict occurred when Russia initiated a full-scale war[124] with the aim of annexing Ukraine, driven by the misguided ambitions of a local dictator who believed that conquering or even destroying the world was within his reach.

In the business world, there are many examples of companies that have not been able to overcome hostility towards each other, which has led to negative results. One example is the rivalry between Blockbuster and Netflix. In the early 2000s, Blockbuster was the dominant player in the video rental market, but Netflix was gaining traction with mail-order DVD subscriptions. Blockbuster pulled out of its partnership with Netflix and instead tried to create its own online video rental service. However, the company's efforts were unsuccessful, and it eventually filed for bankruptcy in 2010.

Another negative example is the rivalry between Uber and Lyft. Both companies emerged in the early 2010s and competed fiercely for market share in the taxi industry. The rivalry between the two companies has led to aggressive tactics, including price wars and driver poaching. This competition ended up hurting both companies, as it led to increased costs and reduced profitability for both Uber and Lyft.

---

[124] Let's clarify: which in 2014 started and in 2022 was actively continued by an illiterate sleepwalker with colossally low intelligence, a retired major dismissed for unsuitability from the post of head of the club, where he followed the manifestation of dissent, a man with monstrous childhood traumas that went right through several generations of his family, was brought to power in a nuclear power by the quirk of a wealthy oligarch, which is an example of the clearly short-sighted behavior of the main adversary of the Soviet Union in the Cold War.

In all these negative examples, the parties involved were unable to overcome their negative emotions and work towards a common goal. This has led to negative consequences, including conflicts, failed partnerships, and reduced profitability. These examples demonstrate the importance of overcoming hostility and finding ways to work together, even in a competitive environment.

It will probably be useful to study the practices of controlling the "reptile brain." What do you think about it?

Principles of Punishment of the Unworthy and Commensurability of Punishment

Punishment for the unworthy must adhere to principles of retribution, deterrence, rehabilitation, and reparation. However, there lurks a risk of abuse within the penitentiary system.

Criminals receiving greater comfort and preferential treatment than in their home communities weaken the preventive force of punishment. To prevent such abuse and preserve the deterrent effect, we ought to consider sending offenders back to their home countries for trial, abiding by local laws and international standards of justice. It is crucial that international law evolves alongside moral progress. In doing so, criminals will not reap the benefits of an advanced system, and their punishments shall align with those in their homelands, restoring the power of deterrence.

Why does this matter now? The world's immigration system is woefully antiquated, unable to keep pace with the speed of communication. Upon reading this book, you will realize that in the event of mass movements of people due to wars or global disasters, the rigid immigration system will mock its proponents' efforts to tighten it. Opponents of immigration raise a fundamental concern — the inconsistency of moral foundations that underpin various legal systems worldwide. Given the rapid development of communication and transportation, continuing to restrain cross-border services will soon be utterly impossible. The sole context that remains for the archaic and foolish notion of closed borders is an increase in crime facilitated by individuals crossing borders, who exploit the host country's lenient penitentiary system. This, in turn, empowers decrepit corrupt officials and their allies, useful idiots, to impede the global development of public morality.

Amidst ongoing technological progress, such obstruction is exceedingly perilous. Morality stands at the forefront of public control over legislation. What prevents a transnational corporation from shaping its own favorable narrative, effectively enslaving people's opinions worldwide for its selfish gain?

Consider, in this regard, the social network Meta (formerly known as Facebook and Instagram). Its initial funding[125] was bestowed by a criminal[126] from an aggressive fascist state, twice convicted of fraud and bribery — a personal confidant of the local dictator. By a remarkable coincidence, these social networks promptly blocked those who protested against the fascist regime, while enabling hired trolls to propagate the regime's agenda. Corrupt leading universities willingly maintain the status quo, provided they can do so discreetly, without causing direct harm to their reputation.

As you can see, the context is immediate and undeniable. In essence, the objective is to uphold the principle of punishment's inevitability and proportionality to the committed crime. This approach is suitable and effective in deterring future criminal behavior.

One solution is to repatriate offenders to their home countries, preventing abuses within the penitentiary system and preserving its deterrent power. If such a solution proves unviable, an alternative is to adopt the norms of local legislation for punishment execution in partner countries. This should be done when the

---

[125] By 2012, Alisher Usmanov's investment in Facebook was $3,000,000,000.

[126] On August 19, 1980, the military tribunal of the Turkestan Military District sentenced Alisher Usmanov, the son of the prosecutor of Tashkent, to 8 years in prison for fraud and complicity in taking a bribe. Together with Usmanov, his friends were convicted — Bahadir Nasymov, son of the deputy chairman of the KGB of Uzbekistan, and Ilham Shaikov, son of the Minister of Agriculture. On March 26, 1986, Usmanov was released on parole due to "sincere repentance" and "exemplary behavior."

punishment imposed in the country where the crime was committed clearly diverges from the offender's moral understanding of appropriate severity. Historical examples, both positive and negative, demonstrate the application of this practice.

Positive historical example.

In the past, certain countries successfully implemented a strategy of repatriating foreign offenders for trial and punishment in accordance with local laws. For instance, in the 1990s, the UK achieved this by forging prisoner transfer agreements with various countries, including Jamaica, Thailand, and India. This approach allowed foreign offenders to serve their sentences in their home countries, alleviating the strain on the UK prison system and reestablishing the deterrent effect of retribution.

OPEN IMMIGRATION ARCHITECTURE.

The United States' history offers numerous examples that underscore the vital role of open immigration in fostering economic prosperity. By upholding the principles of open immigration and implementing effective screening processes, the United States has attracted talented individuals, entrepreneurs, and skilled workers who have made substantial contributions to the nation's economy. Here are key instances:

1. *Immigrant Entrepreneurs and Job Creation*: Throughout American history, immigrants have been pivotal in entrepreneurship and job creation. Many iconic American companies were founded by immigrants or their descendants. Google, Tesla, Intel, and eBay, to name a few, were either founded or co-founded by immigrants or their children. These companies have revolutionized industries,

generating millions of jobs that fuel economic growth and innovation.

2. *Labor Force and Economic Growth*: Immigrants have long been an essential component of the American labor force. They fill gaps in sectors facing shortages, such as agriculture, construction, and hospitality, jobs that native-born workers may be unwilling to undertake. By addressing labor deficiencies, immigrants enhance overall productivity and contribute to the economic growth of the United States.

3. *Skills and Innovation*: Immigrants bring diverse skills, knowledge, and perspectives to the United States. Highly skilled immigrants, including scientists, engineers, and researchers, have made remarkable contributions to technological advancements, medical breakthroughs, and innovation. Since 2000, over one-third of Nobel laureates in the United States have been immigrants. Their contributions not only advance scientific knowledge but also stimulate economic growth and foster competitiveness.

4. *Small Business Ownership*: Immigrants possess a strong entrepreneurial spirit and are more inclined to start small businesses compared to native-born individuals. These enterprises bolster local economies, create jobs, and promote economic vitality. A report by the New American Economy reveals that immigrants are more than twice as likely as native-born Americans to start businesses. Small businesses form the backbone of the U.S. economy, and immigrant entrepreneurs have played a substantial role in their growth and success.

5. *Cultural Diversity and Tourism*: Immigration enriches the United States' cultural tapestry,

rendering it an appealing destination for tourism and international visitors. Immigrants' diverse cultural heritage contributes to the tourism industry, which generates billions of dollars annually. Tourists visit the United States to experience its multicultural communities, cuisine, arts, and traditions — all influenced by immigrants and their descendants.

It is crucial to acknowledge that while open immigration has significantly contributed to the United States' economic prosperity, robust screening processes are imperative to ensure national security and public safety. By implementing effective immigration policies, the United States can strike a balance between embracing immigrants and safeguarding its citizens while reaping the economic benefits of an open and diverse society.

In conclusion, immigrants have made substantial contributions to the United States, bolstering innovation, economic growth, and job creation. Embracing an open immigration system while implementing effective screening processes allows the country to harness the talents, skills, and diverse perspectives that immigrants bring. By doing so, the United States can continue to attract and retain foreign-born talent, ensuring sustained economic prosperity and societal progress.

Negative historical example

However, history has witnessed instances where punishment inflicted on offenders in their home countries was both cruel and degrading, leading to public outcry and condemnation. Take, for instance, the case of Australian journalist Peter Greste, arrested in Egypt in 2013 under charges of disseminating false news and supporting a terrorist organization. In Egypt, he received a seven-year prison sentence, but this verdict faced wide-spread criticism for its alleged political motivations and inherent unfairness. The Greste case exemplifies the utmost importance of fairness in sentencing, not only within one's home country but across all states where the law has been transgressed.

It would be prudent to establish a set of rules that consider these factors for future implementation. Just as in the realm of criminal justice, ensuring fair and equitable business transactions holds paramount significance. One effective approach to curbing system abuse is the establishment of rules and regulations that hold businesses accountable for their actions. These regulations can be applied at international, national, or corporate levels, employing legal sanctions or the potential harm to their reputation as means of enforcement.

In situations where businesses engage in unethical or illegal activities, fines or sanctions commensurate with those imposed in the offender's home country should be implemented. For instance, if a foreign company is found to violate local laws, it should be liable to pay damages or face legal penalties equivalent to fines applicable in its country of origin.

Nevertheless, akin to the principles of criminal justice, there exists a risk of unfairness or disproportionate punishment. It is imperative to ensure that punishment aligns appropriately with the crime committed and does not unjustly burden the innocent. Furthermore, punishment should serve as a deterrent, discouraging unethical or illegal business practices in the future.

Hence, the same principles that govern the realm of criminal justice can be extended to business relationships. Ensuring fair and equitable conduct in business dealings, alongside the appropriate application of fines or sanctions, serves as a crucial deterrent against unethical or illegal activities in the future.

Prominent forms of economic sanctions encompass a range of measures aimed at pressuring or punishing a targeted entity or country. These measures include trade barriers that restrict or regulate the flow of goods and services, asset freezes that seize or block access to financial resources, travel bans that prohibit individuals from entering or leaving a designated area, arms embargoes that prevent the sale or transfer of weaponry, and restrictions on financial transactions that limit the ability to engage in monetary exchanges.

These sanctions are often implemented by governments or international organizations as a means of expressing disapproval, deterring certain behaviors, or seeking to induce policy changes. They can have significant economic and political implications for the targeted entity, impacting its ability to engage in international trade, access financial resources, or interact with the global community.

The effectiveness and consequences of economic sanctions vary depending on various factors, including the targeted entity's resilience, the degree of international support for the sanctions, and the specific measures employed. While sanctions can be a powerful tool in influencing the behavior of individuals, organizations, or countries, they can also have unintended humanitarian consequences and geopolitical ramifications.

Overall, economic sanctions represent a complex and multifaceted approach to shaping international relations, exerting pressure, and promoting desired outcomes through economic means.

The effectiveness of sanctions in achieving their intended goals is a subject of debate. Additionally, the humanitarian impact of country-wide sanctions has been a point of controversy. In recent years, United Nations Security Council (UNSC) sanctions have tended to focus more on individuals and entities rather than comprehensive sanctions against entire countries.

Historically, one notable example of an embargo was the Continental System implemented by Emperor Napoleon I of France during the Napoleonic Wars. The system aimed to cripple the United Kingdom economically but proved to be harmful to all involved parties. Sanctions in the form of blockades were also prominent during World War I, and debates about implementing sanctions through international organizations arose after the war.

The League of Nations permitted the use of sanctions in various cases, including violations of the League Covenant and wars between member states. The Abyssinia Crisis in 1935 resulted in League sanctions against Italy, although they had limited impact. In the lead-up to the Japanese attack on Pearl Harbor in 1941, the United States imposed trade restrictions on Japan to discourage further aggression.

After World War II, the United Nations replaced the League of Nations and the use of sanctions gradually increased during the Cold War. Since the end of the Cold War, there has been a significant rise in economic sanctions. The Global Sanctions Data Base has documented 1,325 sanctions imposed between 1950 and 2022.

Economic sanctions are often used as a foreign policy tool by governments, with objectives ranging from changing policies and resolving territorial conflicts to fighting terrorism and promoting human rights. The effectiveness of economic sanctions can vary depending on the target and the presence of veto players within the government. However, measuring the success of sanctions in achieving their goals is a complex and debated issue.

When nations violate sanctions imposed by international organizations, steps must be taken to prevent the system's abuse and uphold the restraining power of the imposed restrictions. One approach is to employ fines and sanctions comparable to those utilized in the offender's home country.

For instance, if a state is discovered to be breaching international sanctions through engaging in prohibited activities like nuclear proliferation, terrorism, or aiding another nation in such endeavors, the international community may compel it to face legal or economic sanctions equivalent to those imposed within their own country for similar acts. By penalizing all individuals involved in the process and imposing penalties on the assets utilized by such nations, a suitable punishment can be ensured, fostering a deterrent effect against future violations.

Nevertheless, as with criminal justice and business dealings, there exists a risk of unfairness or disproportionate punishment. It is vital to ensure that the punishment aligns appropriately with the offense and does not unjustly infringe upon the legitimate interests of innocent parties. Furthermore, punishment should also function as a deterrent, preventing future transgressions of international sanctions.

In this manner, the same principles governing criminal justice and business relationships can be extended to countries that violate international sanctions. It is imperative to guarantee that fines or sanctions are appropriate, proportionate, and serve as a deterrent against subsequent infractions of international sanctions.

Loyalism in the USA, or the Fifth Column[127]: History and Strategies to Safeguard the Political System

The history of the Fifth Column in the United States offers valuable insights into the country's role as a global enforcer after the Bretton–Woods financial era. By becoming the world's reserve currency, the US dollar

---

[127] A fifth column is any group of people who undermine a larger group or nation from within, usually in favor of an enemy group or another nation. According to Harris Mylonas and Scott Radnitz, "fifth columns" are "domestic actors who work to undermine the national interest, in cooperation with external rivals of the state."

has allowed the nation to borrow money for its military budget at no cost, essentially making it the world's army. The concerns about US debt, while not entirely baseless, are largely irrelevant in this context.

Loyalism during the American Revolutionary War presents a captivating chapter that teaches us important lessons. Loyalists, also known as Tories, Royalists, or King's Men, were colonists who remained faithful to the British Crown. Their presence during that tumultuous period exemplifies both positive and negative aspects of loyalty and allegiance amidst political upheaval. Furthermore, it is crucial to understand the strategies employed to protect the US political system from external forces while upholding the principles of open immigration, which has been instrumental in America's success.

During the late 1700s, loyalists found themselves on the opposing side as the Patriots fought for independence. Many loyalists believed in the legitimacy of the British government and considered rebellion against the Crown a betrayal to the Empire. These loyalists, often older and established individuals, resisted radical change and feared the potential chaos and corruption that could accompany revolution. They valued order and recognized Parliament as the rightful authority. The concept of separate sovereign states was revolutionary, as the American identity was still taking shape.

On the other hand, the Patriots, seeking independence and self-governance, viewed loyalists with suspicion. They saw them as a threat to their cause and actively suppressed any organized loyalist opposition. The Patriots' victory in the war resulted in the expulsion or flight of many loyalists from the Thirteen Colonies.

Approximately 15 percent of loyalists, numbering between 65,000 and 70,000 individuals, sought refuge in other parts of the British Empire, including Britain itself and British North America (now Canada). Southern loyalists migrated to Florida, which remained loyal to the Crown, along with the British Caribbean possessions. Northern loyalists predominantly settled in Ontario, Quebec, New Brunswick, and Nova Scotia, forming a group known as the United Empire Loyalists. Although many loyalists eventually returned to the United States, discriminatory laws had to be repealed to ensure their full integration into American society.

The history of loyalism sheds light on the dynamics of loyalty, allegiance, and the challenges faced by those who remain faithful to an oppressive or foreign government in a new nation. It underscores the importance of recognizing diverse motivations that drive individuals to hold differing opinions and loyalties, particularly in the face of conflicting political views.

Transposing this historical context to the present, the United States faces the unique challenge of protecting its political system from external disruptions while maintaining an open immigration infrastructure. Immigration has been a fundamental aspect of the nation's success, attracting individuals from diverse backgrounds who contribute to its vibrant tapestry. However, it is crucial to remain vigilant against potential threats that may exploit this openness.

To safeguard the political system, history offers valuable guidance. First and foremost, maintaining a robust and impartial judicial system is crucial. The rule of law must prevail, ensuring equal treatment for all individuals, regardless of their background or political beliefs. This ensures that external forces or radical elements cannot undermine the democratic process through unlawful activities.

Encouraging civic engagement and responsible citizenship strengthens the political system against external disruptions. Active participation in the democratic process fosters a sense of investment in the nation's well-being, making individuals more likely to reject extremist ideologies that undermine democratic principles.

Promoting open and informed dialogue is essential. Respectful and constructive discussions allow individuals to voice their concerns, express grievances, and propose solutions. This approach counters divisive narratives and prevents the polarization that can be exploited by external actors seeking to sow discord.

Finally, maintaining strong border security measures while upholding the principles of open immigration is crucial. Adherence to established immigration processes allows for effective screening of potential threats without compromising the commitment to welcoming immigrants.

In conclusion, the lessons from loyalism during the American Revolutionary War continue to resonate today. They reveal the complexities of loyalty, the diverse motivations driving allegiances, and the challenges faced by individuals caught between conflicting political views. By understanding these lessons, the United States can effectively protect its political system from external disruptions while upholding its commitment to open immigration, which has been integral to the nation's success. Through an impartial judiciary, civic engagement, open dialogue, and robust border security measures, the United States can preserve its democratic values and ensure a prosperous future for all its citizens.

The risk materialization takes various forms. While open immigration infrastructure is crucial for the prosperity of US society, it is essential to recognize the potential dangers posed by unintegrated immigrants, including criminal activities, unethical business practices, and support of foreign states, which can be considered treasonous. In recent years, countries like China and Russia have made known efforts to manipulate the US political system, including interference in the 2016 and 2020 elections. It is also important to acknowledge that Russian agents have employed Russian-speaking declassified elements to promote narratives favoring Russia on the internet. Furthermore, Russia has financed various "cultural venues" such as movies, Russian military parades in the USA, radio shows, and advertisements across different media to propagate pro-Russian propaganda.

Russian Interference

The Russian government, through entities like the Internet Research Agency, has been known to manipulate public opinion and spread disinformation during the 2016 United States presidential election. They have used not only trolls but also Russian-speaking declassified elements to promote narratives that favor Russia on the internet. Additionally, Russia has financed cultural venues, including pro-Russian propaganda movies, Russian military parades in the USA, and radio shows, to propagate their desired narrative. These activities aim to shape public perception, promote a pro-Russian agenda, and influence political outcomes.

Chinese Influence

China has also been involved in shaping political discourse and interfering in the United States. While unrelated to the previously mentioned cultural venues, China has employed various strategies to exert influence, including economic coercion, intellectual property theft, and disinformation campaigns. These activities pose significant economic and national security risks that necessitate careful monitoring and countermeasures.

Other Examples

It is important to note that efforts to manipulate public opinion and promote specific narratives extend beyond

Russia and China. Various countries, including Albania, Brazil, India, North Macedonia, Malaysia, the Philippines, Turkey, and Nicaragua, have utilized troll farms or similar operations to shape public opinion through propaganda, disinformation, and misinformation. Financing cultural venues and media outlets to promote specific ideologies or narratives is not exclusive to Russia but is also practiced by other countries seeking to influence public perception.

Recognizing and addressing these threats is crucial to safeguarding the integrity of the US political system. Robust cybersecurity measures, intelligence gathering, and counterintelligence efforts are necessary to identify and mitigate foreign interference attempts. Additionally, promoting digital literacy, media literacy, and critical thinking skills among the population can help individuals better discern between genuine information and propaganda.

Maintaining an open immigration system while ensuring effective screening processes is essential to strike a balance between welcoming immigrants and protecting national security. Thorough vetting and background checks aid in identifying individuals with potential ties to foreign actors or involvement in criminal activities. It is important to emphasize that these measures should not be used to stigmatize or discriminate against immigrants as a whole, but rather to ensure the safety and well-being of the nation.

In conclusion, while open immigration infrastructure is crucial for the prosperity of the United States, it is equally important to address the potential risks associated with unintegrated immigrants, including criminal activities and support of foreign states. Recognizing and countering foreign interference efforts, as observed in recent elections, is vital to safeguard the democratic processes and institutions of the country. Striking a balance between openness and security is key to ensuring a prosperous and resilient society, while remaining vigilant against attempts to manipulate public opinion and influence national affairs.

Specific instances of Russian activity are evident. Aleksandr Ionov, a Russian national, stands accused of conspiring with U.S. citizens to act as illicit agents of the Russian government. Ionov, working in tandem with the Russian Federal Security Service (FSB), orchestrated a long term campaign aimed at sowing discord, disseminating pro-Russian propaganda, and interfering in U.S. elections.

The indictment reveals that Ionov, along with at least three Russian officials, embarked on this campaign spanning from December 2014 to March 2022. Ionov, the founder and president of the Anti-Globalization Movement of Russia (AGMR), utilized this organization, funded by the Russian government, to execute Russia's manipulative campaign.

Chinese interference has been on the rise, as China intensifies its efforts to influence U.S. politics, media, and society. Their tactics involve manipulating narratives, promoting politicians sympathetic to Beijing, and spreading chaos and falsehoods. This pattern of influence operations initially emerged in the Pacific Rim but has now expanded its reach to the United States.

In countries like Australia, China-linked donors have directly influenced politicians through financial means, aiming to shape foreign policy in favor of China's interests. Additionally, China has supported pro-Beijing business figures in acquiring control over local Chinese-language media outlets in regions such as New Zealand, Taiwan, and Southeast Asia.

Former President Donald Trump often portrayed China as a strategic competitor and, at times, even as a foe of the United States. To fully grasp the relationship between the Trump administration and China, it is crucial to examine the actions and policies implemented during his presidency.

Trade Policy

Trump's actions raised doubts about his stance on China, particularly in trade policy. In 2018, the United States entered into a trade war with China, imposing tariffs on various Chinese goods and initiating negotiations for a new trade agreement. The objective was to address long-standing issues such as intellectual property theft, forced technology transfer, and unfair trade practices. While these actions aimed to rebalance the trade relationship, some argued that they ultimately favored China. The tariffs imposed by the Trump administration resulted in increased costs for American businesses and consumers, while China sought alternative markets and expanded its global influence.

Diplomatic Engagement

Another aspect worth considering is diplomatic engagement. Despite the tough rhetoric, Trump pursued several initiatives that appeared to prioritize cooperation with China. During his presidency, Trump made efforts to engage with Chinese President Xi Jinping on multiple occasions, expressing a willingness to negotiate and resolve trade disputes. There were instances where Trump spoke positively about his personal relationship with President Xi, indicating a desire for productive engagement.

COVID-19 Pandemic

The handling of the COVID-19 pandemic also shed light on the complexities of the Trump administration's approach to China. While Trump criticized China for its management of the outbreak and raised transparency concerns, there were moments when he acknowledged China's efforts to control the virus. However, his administration faced criticism for downplaying the severity of the pandemic in its early stages and for its own response to the crisis.

Chinese Interference

China's efforts to influence U.S. politics, media, and society have extended beyond its immediate vicinity. Chinese state media organizations, such as CGTN, have invested significant resources in influencing U.S. politics. Private media outlets sympathetic to Beijing dominate the Chinese-language television, print, and online media landscape. These channels serve as platforms for propagating Chinese propaganda, potentially influencing voting behavior, especially in competitive congressional districts.

Beijing has also employed deceptive tactics, including undisclosed advertorials and the dissemination of propaganda through Chinese Radio International on U.S. radio stations. Unlike Russia, which often seeks to create chaos, China aims to reshape broader perceptions of itself.

The FBI has reported a notable increase in investigations related to China, and the U.S. National Counter-intelligence and Security Center has warned of China's targeting of local politicians, mayors, governors, and state legislators to influence national politics in various ways. There have been cases where suspected Chinese spies infiltrated political circles, as seen with Fang Fang in California's Bay Area.

China has escalated its use of digital disinformation to interfere in U.S. elections. By concealing the origins of disinformation through bots and fake profiles, China attempts to avoid its negative global image. Chinese tactics mirror those employed by Russia, fostering online animosity among Americans regarding divisive issues.

The United States has responded by intensifying scrutiny of China but faces challenges in building cases against Chinese nationals and Chinese Americans. The FBI and the State Department's Global Engagement Center have improved their abilities to detect Chinese disinformation online, and major tech companies have made efforts to label state media and remove dubious posts.

Regarding espionage, the FBI needs to broaden its focus on local politics as China seeks to cultivate relationships with lower-ranking officials, aiming to obtain information and exert influence if these officials ascend to higher positions or participate in major elections.

The FBI has warned of potential interference in the 2022 elections, primarily through China's political influence campaigns, control of Chinese-language media, and online disinformation. Chinese influence in U.S. politics, media, and society is expected to continue expanding in the years ahead.

A Noteworthy Case in NYC

In addition to China's broader efforts to influence U.S. politics, media, and society, two individuals were apprehended in New York City for operating an illegal overseas police station on behalf of the Ministry of Public Security of the People's Republic of China (PRC). Chen Jinping faced charges of acting as agents of the PRC government and obstructing justice by destroying evidence of their communications with an MPS official.

Chen collaborated with a number of people in the USA to establish the first overseas police station in the United States for the Fuzhou branch of the MPS. The secret police station, located in Manhattan's Chinatown, ceased operations in 2022 after the individuals involved became aware of the FBI's investigation. They chose not to inform the U.S. government about their involvement in operating an illicit MPS police station on U.S. soil.

The establishment of the police station violated U.S. sovereignty, granting the PRC the ability to monitor and intimidate dissidents critical of its government. The actions of the PRC and its security apparatus surpass acceptable conduct for a nation-state. While the U.S. government is committed to safeguarding the freedoms of all individuals residing in the country from the threat of authoritarian repression, the very fact of PRC acting so agressively is alarming.

The individuals involved in running the police station were tasked with aiding the PRC government's repressive activities on U.S. soil. Their actions included participating in counterprotests against a prohibited religion during President Xi Jinping's visit in 2015, attempting to coerce a supposed fugitive to return to the PRC in 2018, and assisting in locating a pro-democracy activist in California in 2022.

The episode represents an example of how governments, in this case, the United States, acted against perceived threats from radical movements. The deportation of individuals associated with aggressive rebellion and anarchist ideologies aimed to preserve social order and safeguard the existing political system.

Possible actions

To prevent the destabilization of the U.S. political system and safeguard against clandestine foreign interference, a range of measures can be taken. Here are some potential actions to ensure the safety and integrity of the United States:

1. *Strengthen Cybersecurity*: Bolster defenses and intelligence capabilities to protect critical infrastructure, government networks, and electoral systems from cyber threats and disinformation campaigns.

2. *Enhance Counterintelligence Efforts*: Heighten monitoring and investigations of suspicious

activities by foreign agents, fostering collaboration between intelligence agencies and sharing information with key allies.

3. *Enforce Foreign Agents Registration Act (FARA)*: Vigorously ensure transparency and accountability in foreign influence on U.S. politics by monitoring lobbying activities, funding sources, and interactions between foreign entities and U.S. politicians.

4. *Enhance Electoral Security*: Implement robust safeguards such as secure voter registration systems, paper trail auditability, and measures to prevent foreign interference in election campaigns, while strengthening cooperation among federal, state, and local authorities.

5. *Raise Public Awareness*: Promote media literacy and critical thinking skills to educate the public about foreign influence operations, disinformation campaigns, and propaganda efforts, enabling citizens to identify and counter false narratives.

6. *Support Democratic Institutions*: Invest in strengthening democratic institutions, civil society organizations, and independent media to maintain resilience against foreign attempts to undermine them, ensuring transparency, accountability, and public trust.

7. *International Cooperation*: Collaborate closely with like-minded democracies to share information, coordinate efforts, and develop joint strategies to counter foreign interference, forming alliances and partnerships to collectively defend against influence campaigns.

8. *Diplomatic Engagement*: Engage in diplomatic efforts to address and deter foreign interference, raising concerns with foreign governments,

imposing diplomatic consequences for malicious activities, and advocating for international norms against interference.

9. *Robust Intelligence Gathering*: Invest in intelligence capabilities to gather information on foreign interference activities and anticipate potential threats, enhancing intelligence sharing and cooperation among agencies.

10. *Legislative Reforms*: Review and update existing laws to address emerging threats and challenges posed by foreign interference, considering measures that increase transparency, enhance accountability, and impose consequences for illicit activities.

Refugee Contribution

Refugees can play a role in safeguarding the U.S. political system by helping to identify and expose Fifth Column agents, including Russian and Chinese agents. Here's how refugees can contribute while ensuring U.S. safety:

1. *Intelligence Sharing*: Collaborate closely with intelligence agencies to identify and expose foreign agents and their networks within their diaspora, assisting in countering foreign interference.

2. *Counterintelligence Operations*: Enhance counterintelligence efforts to detect and disrupt foreign intelligence operations targeting the United States, actively investigating and exposing covert activities.

3. *Cybersecurity Collaboration*: Share expertise and collaborate in cybersecurity with the countries where refugees have sought refuge, strengthening defenses against foreign cyber threats.

4. *Support in Diplomatic Efforts*: Support diplomatic efforts to counter foreign agency influence by

advocating for international norms against interference and actively participating in international forums and organizations.

5. *Sharing Lessons Learned*: Share firsthand experiences dealing with foreign interference and disinformation campaigns, contributing to the improvement of U.S. responses and resilience against foreign influence.

6. *Support for Democratic Institutions*: Support U.S. initiatives aimed at strengthening democratic institutions, civil society organizations, and independent media, sharing experiences and best practices to safeguard against foreign interference.

7. *Assistance in Investigations*: Provide information, evidence, and cooperation in investigations related to foreign interference, aiding in legal proceedings to hold those involved in covert activities accountable.

8. *Promote Public Awareness*: Contribute to international efforts to raise public awareness about foreign interference, disinformation campaigns, and propaganda, sharing experiences and supporting initiatives promoting media literacy and critical thinking.

By collaborating with the United States and actively participating in efforts to combat foreign interference, refugees can demonstrate their commitment to shared democratic values, contributing to the safety and integrity of the U.S. political system. This partnership can strengthen the resilience of both countries against foreign malign influence, ensuring the integrity of democratic processes.

Collaboration Potential

The opposition leaders in Russia and China possess the capacity to aid in the identification of Russian and

Chinese agents and illegal representatives in the United States. They can achieve this through the following means:

1. *Insider Knowledge*: Opposition leaders often hold insights and information regarding the activities of their respective governments. They may have connections within intelligence agencies or government institutions, allowing them to acquire valuable information on foreign agents and their operations.

2. *Whistleblower Support*: Opposition leaders can encourage individuals with knowledge of foreign espionage activities to step forward as whistleblowers. By providing platforms for reporting and ensuring the protection of whistle-blowers, opposition leaders facilitate the disclosure of crucial information on foreign agents.

3. *International Advocacy*: Opposition leaders can raise awareness on foreign interference and espionage by advocating for stricter measures and increased transparency. They can collaborate with international organizations, human rights groups, and governments to shed light on the activities of foreign agents and push for stronger countermeasures.

4. *Cooperation with Intelligence Agencies*: Opposition leaders can collaborate with intelligence agencies from other countries, including the United States, to exchange information and intelligence on foreign agents. Such cooperation helps identify individuals engaged in espionage activities and expose their networks.

5. *Lobbying Efforts*: Opposition leaders can engage in lobbying endeavors aimed at influencing policies and legislation pertaining to foreign interference and espionage. Through collaboration with

lawmakers and government officials, they can advocate for stricter regulations and enforcement measures to counter the activities of foreign agents.

6. *Media Exposure*: Opposition leaders can utilize media platforms to expose and raise public awareness about the presence and activities of foreign agents. By highlighting specific cases, providing evidence, and disseminating information, they contribute to the identification and exposure of foreign agents.

It is important to recognize that the contributions of opposition leaders in identifying foreign agents may encounter challenges and risks. They may face obstacles such as government repression or limited access to information. Furthermore, their involvement should align with legal frameworks and be coordinated with relevant authorities to ensure the accuracy and legitimacy of the information provided.

Sanctions Evasion

The patterns of evading sanctions pose dangers to the world. While reports of illicit activities involving the circumvention of U.S. sanctions on Iran exist, it is important to note that discussing specific schemes or allegations without concrete evidence can be speculative. However, I can offer some general information on sanctions evasion and the potential role of Syria, Russia, China, and Turkey:

1. *Sanctions Evasion*: Sanctions evasion encompasses various techniques employed to bypass or undermine economic sanctions imposed on a country. These techniques include illicit financial transactions, smuggling, mislabeling of goods, and the utilization of front companies or intermediaries to obscure the true origin or destination of goods.

2. *Syria's Role*: Due to Syria's strategic location and ongoing conflict, concerns have been raised regarding potential sanctions evasion activities within its borders. Syria has historically maintained close ties with Iran, and it is plausible that entities or individuals may seek to exploit these connections to facilitate illicit trade or circumvent sanctions.

3. *Russia and China's Influence*: Russia and China, as significant global powers, possess economic and political relationships with various countries, including Iran and Syria. Although concrete evidence is necessary to establish their involvement in specific sanctions evasion schemes, their support for these countries and their interests in the region could potentially enable or facilitate activities that undermine sanctions.

4. *Turkey's Involvement*: Turkey, as a neighboring country to both Syria and Iran, plays a significant role in regional dynamics. While Turkey is a NATO ally of the United States, differing objectives and strategic considerations have occasionally strained bilateral relations. Reports and allegations of sanctions evasion involving Turkish entities exist, but specific details and evidence are required to draw definitive conclusions.

Understanding that addressing sanctions evasion and illicit activities necessitates comprehensive intelligence gathering, international cooperation, and legal mechanisms is crucial. Governments, including the United States, closely monitor and investigate potential violations of sanctions to mitigate the risk of evasion and maintain the effectiveness of these economic measures. Investigating and uncovering specific schemes or activities related to sanctions evasion requires robust intelligence capabilities, international cooperation, and thorough analysis of evidence.

The Next Threats for the USA

The subsequent threats following the absorption of the situation in Ukraine, likely including the deployment and usage of nuclear weapons by Russia, originate from the Middle East. The threats to the United States from the Middle East, potentially inspired or influenced by Russia and China, as well as the circumvention of sanctions by Russia and Iran, present complex and multifaceted challenges. Here is a brief elaboration on these issues:

1. *Iran*: Iran presents various challenges to the United States and its interests in the Middle East. It stands accused of supporting and financing terrorist organizations such as Hezbollah and Hamas, which pose threats to regional stability and U.S. allies. Concerns arise from Iran's pursuit of nuclear weapons and its ballistic missile program, raising questions about its intentions and capabilities. Regarding sanctions circumvention, Iran has sought ways to evade international sanctions imposed on its nuclear program and other activities. This includes utilizing sophisticated financial mechanisms, engaging in illicit trade networks, and employing front companies to obfuscate the true origin of goods and funds.

2. *Syria*: The ongoing conflict in Syria has created a complex security situation in the region. While Russia has significantly supported the Syrian regime, China has also displayed some level of support. Russia and China's involvement in Syria has allowed them to expand their influence and secure economic and strategic interests in the region. Within this context, extremist groups such as ISIS, which originated from the efforts of the KGB during the Soviet Union era, and other jihadist organizations have exploited the power

vacuum and instability in Syria. Russia applies every effort to destabilize the situation in Syria and adjacent territories to help Iran smuggle crude oil to the illegal private oil refineries located on the border of Syria and Turkey on the territory of Kurdistan — a self-proclaimed autonomy of Kurds in Turkey. Later the refined oil products are mixed with the oil in the oil tanker vessels at the Mediterranean ports. These groups pose security threats not only to the United States but also to its allies and the democratic world as a whole, as they seek to establish a foothold and propagate their extremist ideologies.

Sanctions Circumvention

Both Russia and Iran have endeavored to circumvent international sanctions imposed on them by employing various strategies. These strategies encompass the utilization of front companies, conducting illicit financial transactions, engaging in trade-based money laundering, and participating in barter trade to evade detection and sustain their illicit activities.

The Russian-speaking populations in countries like Georgia, Armenia, Uzbekistan, and Belarus may possess firsthand knowledge or connections that can aid in identifying individuals involved in sanctions evasion or anti-American activities. Their understanding of local networks, languages, and cultural dynamics holds value in the identification and exposure of these activities.

It is important to note that addressing these threats requires a multifaceted approach, including diplomatic efforts, intelligence cooperation, economic measures, and targeted counterterrorism operations. Collaborating with regional partners and leveraging the expertise of individuals familiar with local dynamics enhances the

United States' ability to mitigate the threats emanating from the Middle East and counter the activities of Russia, China, and their proxies.

In summary, in addition to the discussion on sanctions evasion and the dynamics of the Middle East, it is crucial to timely address Russia and China's interference in the United States and the challenges posed by climate change:

1. *Foreign Interference*: Countries such as Russia and China have been known to engage in various forms of interference and influence operations targeting the United States. These activities encompass cyberattacks, disinformation campaigns, and attempts to manipulate public opinion, political processes, and social divisions. Such interference carries significant implications for national security, democratic processes, and international relations.

2. *Climate Change and Population Relocation*: Climate change presents a substantial global challenge that will reshape societies and populations. Its impact is expected to result in rising sea levels, extreme weather events, changes in agricultural patterns, and other consequences. These factors contribute to population displacement and migration on an unprecedented scale, with an estimated 226 million people potentially displaced by 2050 due to climate change. Relocating such a significant number of people entails substantial social, economic, and political implications, both domestically and globally. Addressing this issue necessitates robust international cooperation, humanitarian assistance, and comprehensive policies to meet the needs of those affected and ensure their well-being, security, and integration.

To tackle these challenges, governments, international organizations, and civil society must collaborate to:

- Strengthen cybersecurity measures and resilience to counter foreign interference and protect democratic processes.
- Enhance intelligence capabilities and information sharing to identify and respond to disinformation campaigns.
- Foster international cooperation and multilateral partnerships to address climate change, mitigate its impacts, and support vulnerable populations.
- Implement policies and investments that promote climate resilience, sustainable development, and the transition to a low-carbon economy.
- Strengthen humanitarian response mechanisms and provide assistance to communities affected by climate-induced displacement.
- Foster global dialogue and cooperation to develop comprehensive solutions for climate-related population relocation, including resettlement, adaptation strategies, and support for affected communities.

The challenges posed by foreign interference and climate change necessitate sustained international attention, collaborative efforts, and innovative solutions. By comprehensively addressing these issues, the global community can strive to safeguard national security, protect democratic processes, mitigate the impacts of climate change, and uphold human rights while promoting sustainable development.

Consider alternative methods to enable the free trade and immigration with appropriate safeguards.

Chapter XIX

Principles of Encouragement of Those Who Have Stumbled to Correction

The principle of encouraging the correction of those who have stumbled is rooted in the belief that individuals possess the capacity for change and can be motivated to improve their behavior. This principle holds particular relevance in cases where individuals have committed wrongful acts or errors and have faced conviction for their actions. Instead of simply imposing punishment, the aim is to instill encouragement in them, leading them to acknowledge their mistakes, take responsibility for their actions, and actively strive to rectify them.

Positive instances of applying this principle can be observed in restorative justice programs that focus on rehabilitating offenders and addressing the harm caused by their actions. These programs encourage offenders to acknowledge their actions, make reparations to their victims, and engage in activities that foster the development of skills and values promoting socially responsible behavior. For instance, certain restorative justice programs incorporate community service or counseling, which aid offenders in acquiring new skills, cultivating empathy, and fostering a sense of purpose and self-esteem.

Additionally, the principle of encouraging the correction of those who have stumbled finds application in the field of addiction treatment. Numerous addiction treatment programs center around helping individuals recognize the detrimental impact of their addictive behaviors, assume responsibility for their actions, and develop effective coping strategies. Such programs typically encompass a combination of therapy, medication, and participation in support groups, which collectively assist individuals in addressing the underlying psychological and social factors contributing to their addiction.

Conversely, negative examples of this principle manifest in situations where individuals who have made mistakes are stigmatized or subjected to shame for their actions. In such cases, individuals may become disinclined to seek help or initiate positive changes due to the fear of being judged or ostracized by society. This is particularly devastating in instances where individuals are already grappling with mental health issues or addiction, as they may already experience feelings of isolation and stigmatization.

Thus, the principle of encouraging the correction of those who have stumbled rests upon the understanding that people possess the capacity for change and can be motivated to improve their behavior. Positive illustrations of this principle can be observed in restorative justice and drug treatment programs, whereas negative examples arise when individuals are stigmatized or shamed for their mistakes.

Within the principle of encouraging the correction of those who have stumbled, it becomes crucial to consider the criteria for determining suitability for correction, as well as the inclusion of those who may opt not to partake in the correctional process. The objective is to ensure that the remediation process effectively promotes positive change while also addressing the needs of those who have stumbled and are willing to accept assistance.

Generally, eligibility criteria for correctional programs may vary depending on the specific context. Some common factors to consider include the severity of the offense, the offender's motivation to change, and the availability of resources for restitution. For instance, within the realm of criminal justice, eligibility for correctional programs may hinge on elements such as the nature of the crime, the offender's criminal record, and the availability of resources for rehabilitation.

Another crucial consideration pertains to the individual's willingness to participate in the correctional process. While it is important to encourage individuals to seek help and strive to improve their behavior, it is equally vital to respect their autonomy and their right to choose whether or not to engage in correctional programs. In instances where an individual elects not to participate, it may become necessary to explore alternative forms of intervention, such as enhanced monitoring or the application of consequences for ongoing problematic behaviors.

Moreover, it is essential to acknowledge that certain individuals may face barriers when it comes to participating in correctional programs, such as financial hardships, limited access to resources, or cultural and language barriers. The elimination of these barriers may be imperative to ensure that correctional programs are accessible and effective for all eligible individuals.

Therefore, eligibility criteria for correctional programs may depend on factors such as the severity of the offense and the offender's motivation to change. Nonetheless, it is crucial to respect the autonomy of individuals and their right to choose whether or not to participate in correctional programs. Removing barriers to participation may also prove necessary to guarantee the accessibility and effectiveness of correctional programs for all eligible individuals.

Numerous alternative approaches to traditional criminal correction have been proposed and implemented across various contexts. Some of these alternatives have gained relative prominence and have been utilized in diverse forms over several years, while others are more experimental or unconventional. The following examples encompass alternative approaches to criminal correction, including some that may be considered unconventional:

Restorative justice

Restorative justice represents an approach to criminal justice that aims to address the harm caused by criminal behavior, rather than solely focusing on punishing the offender. This approach often involves dialogue between the offender, the victim, and community members, and may include reparations, such as community service or financial compensation. Restorative justice has been implemented in a wide range of contexts, ranging from local community programs to national criminal justice systems.

Therapeutic jurisprudence

Therapeutic jurisprudence embodies an approach to the practice of law that prioritizes the psychological and emotional well-being of individuals involved in legal proceedings. This may involve employing psychological evaluations to inform sentencing or implementing alternative programs that provide treatment for underlying mental health or substance abuse issues. Therapeutic jurisprudence has found application in various contexts, including drug courts and mental health courts.

Animal therapy

Animal therapy encompasses the use of trained animals to facilitate therapeutic interventions with individuals who have been involved in criminal behavior. This approach may entail working with dogs, horses, or other animals to assist individuals in developing empathy, reducing anxiety, or cultivating other socially responsible behaviors. Animal therapy has been employed in diverse settings, ranging from prisons to juvenile correctional facilities.

Rehabilitation in virtual reality

Virtual reality rehabilitation entails employing immersive[128] technology to create simulations of real-world environments, aiding individuals in developing new skills

---

[128] *Immersive technologies* imply the transformation of the role of the teacher, focusing on the design of a multimodal virtual environment, creating immersion scenarios. Over the past few years, "immersiveness" in education has been recognized as a powerful and effective tool to support learning.

or overcoming psychological barriers. This approach has been utilized in various contexts, including helping individuals with anxiety disorders or post-traumatic stress disorder. It has also been proposed as a tool to help individuals involved in criminal behavior develop empathy or other socially responsible behaviors.

Ecotherapy

Ecotherapy involves utilizing natural interventions to enhance emotional and psychological well-being. This may include engaging in activities like gardening, hiking, or other outdoor endeavors that enable individuals to connect with nature and develop a sense of purpose and meaning. Ecotherapy has been implemented in diverse contexts, including prisons and juvenile detention centers.

Art Therapy

Art therapy utilizes creative activities such as painting, drawing, or sculpting to improve emotional and psychological well-being. This approach has been employed across a wide array of settings, including prisons and juvenile detention centers, and has proven effective in assisting individuals in developing new skills and coping strategies.

While some of these approaches may be considered unconventional or exotic, they have all demonstrated effectiveness in promoting positive change among individuals involved in criminal behavior. It is crucial to continue exploring and developing new approaches to criminal correction, as traditional punitive approaches have proven ineffective in reducing recidivism and fostering long term positive change.

To delve further into the influence and philosophical themes of rehabilitation, we inevitably encounter the principles of free choice and the context surrounding them. This concept has been extensively explored in a movie called *The Matrix*[129], with its connection to Plato's *Allegory of the Cave*[130]. The film follows the story of Neo, who discovers the truth about the world he inhabits and chooses to perceive the reality outside the simulated Matrix. The journey out of the cave in Plato's allegory is compared to Neo's awakening from the artificial reality. The text delves into the concept of whether human beings genuinely desire the truth and explores the resistance and denial often accompanying the revelation of truth. It also examines the notion of finding solace in ignorance and the suspension of disbelief when confronted with unreal or

---

[129] The Matrix is a 1999 science fiction action film written and directed by the Wachowskis. It is the first installment in the Matrix film series, starring Keanu Reeves, Laurence Fishburne, Carrie-Anne Moss, Hugo Weaving, and Joe Pantoliano, and depicts a dystopian future in which humanity is unknowingly trapped inside the Matrix, a simulated reality that intelligent machines have created to distract humans while using their bodies as an energy source.

[130] *The Allegory of the Cave*, or *Plato's Cave*, is an allegory presented by the Greek philosopher Plato in his work Republic (514a–520a) to compare "the effect of education (παιδεία) and the lack of it on our nature". It is written as a dialogue between Plato's brother Glaucon and his mentor Socrates, narrated by the latter.

In the allegory "The Cave", Plato describes a group of people who have lived chained to the wall of a cave all their lives, facing a blank wall. The people watch shadows projected on the wall from objects passing in front of a fire behind them and give names to these shadows. The shadows are the prisoners' reality but are not accurate representations of the real world. The shadows represent the fragment of reality that we can normally perceive through our senses, while the objects under the sun represent the true forms of objects that we can only perceive through reason. Three higher levels exist: the natural sciences; mathematics, geometry, and deductive logic; and the theory of forms.

Socrates explains how the philosopher is like a prisoner who is freed from the cave and comes to understand that the shadows on the wall are not the direct source of the images seen. A philosopher aims to understand and perceive the higher levels of reality. However, the other inmates of the cave do not even desire to leave their prison, for they know no better life

illusory experiences. The character Cypher is analyzed as someone who rejects the truth and instead prefers a false reality due to its comforts. This leads us to the potential conclusion that perhaps a significant portion of humanity does not genuinely seek the truth, while the value of isolating humans behind the veil of an artificial reality could be seen as a reasonably fair solution, especially considering humanity's continuous attempts to destroy our magnificent planet. Only a few individuals actively seek a meaningful narrative and a sense of belonging.

It might be worthwhile for all of us to carefully contemplate these ideas.

Principles of Charity
and Giving

C harity plays a vital role in the development of the universe by fostering compassion, promoting social cohesion, and addressing systemic inequalities.

Here are some specific reasons why I think charity is important for the development of the universe:

1. *Alleviating suffering*: Charity helps to alleviate the suffering of individuals and communities facing various hardships such as poverty, hunger, homelessness, and illness. By providing resources, support, and assistance to those in need, charity contributes to the well-being and overall development of individuals, allowing them to lead more fulfilling lives.

2. *Promoting social justice*: Charity endeavors often focus on addressing systemic issues and promoting social justice. By addressing root causes of inequality and injustice, such as lack of access to education, healthcare, and basic needs, charity aims to create a fairer society where everyone has equal opportunities for development and success.

3. *Fostering empathy and compassion*: Engaging in charitable acts cultivates empathy and compassion among individuals. It encourages people to understand the challenges and struggles faced by others, promoting a sense of interconnectedness and unity within the universe. This fosters a more caring and supportive society that values the well-being of all its members.

4. *Creating positive change*: Charitable organizations and initiatives have the power to create positive change on a large scale. Through their work, they can advocate for policy reforms, raise awareness about pressing social issues, and mobilize resources

to address complex challenges. This collective effort contributes to the development and progress of societies, promoting a more equitable and sustainable world.

5. *Encouraging collaboration and community involvement*: Charity often brings together individuals, organizations, and communities to work towards a common goal. This collaboration fosters a sense of community involvement and active participation in the betterment of society. It strengthens social bonds and encourages individuals to take collective action for positive change.

6. *Inspiring others*: Charitable acts have the potential to inspire others to get involved and make a difference. When individuals witness the positive impact of charitable efforts, it can motivate them to contribute their time, resources, and skills to causes they care about. This ripple effect leads to a broader culture of giving and philanthropy, further enhancing the development and well-being of the universe.

In summary, charity is essential for the development of the universe as it addresses immediate needs, promotes social justice, cultivates empathy, creates positive change, encourages collaboration, and inspires others to make a difference. All this is the path to the creation of the positive emotions and therefore by actively engaging in charitable acts, individuals and communities contribute to a more compassionate, equitable, and flourishing universe.

When we engage in acts of giving, such as financial donations, volunteering, or donating goods, we extend help to those in need and make a positive impact on their lives. However, the benefits of charity extend beyond the recipients themselves. Engaging in charitable activities also has several positive effects on our own lives.

Firstly, participating in charity gives us a sense of purpose and fulfillment. Knowing that we are contributing to the well-being of others and acting for the greater good can bring a deep sense of satisfaction and meaning to our lives. It allows us to connect with something larger than ourselves and make a tangible difference in the world.

Studies have shown that engaging in charitable acts can also increase happiness and improve mental health. When we give to others, it activates the reward centers in our brains and releases feel-good chemicals like dopamine. This can contribute to a greater overall sense of happiness and well-being. Additionally, acts of kindness and generosity have been linked to lower stress levels and improved mental health outcomes.

Furthermore, charity presents an opportunity for personal growth and development. By giving to others, we develop important qualities such as empathy, compassion, and a deeper understanding of the needs and challenges faced by others. These qualities not only strengthen our relationships with others but also allow us to navigate the world with greater empathy and understanding.

In addition, engaging in philanthropy and community involvement can have positive implications for our professional lives. Actively participating in charitable activities can help build a positive reputation, both personally and within our communities or industries. It showcases qualities such as leadership, empathy, and a desire to make a positive impact, which are highly valued by employers and peers. Furthermore, engaging in philanthropic endeavors can open up networking opportunities and expand our connections with like-minded individuals and organizations.

Overall, giving and charity are important for personal and community development. They have the power to bring about positive changes in individuals and society as a whole. By giving to others, we experience a sense of accomplishment and fulfillment, develop essential qualities like empathy and compassion, and enhance our mental and physical well-being. Therefore, incorporating philanthropy into our lives can be immensely beneficial, not only for the recipients of our generosity but also for our own personal growth and well-being.

> In the Kabbalistic tradition, there are eight internal principles of philanthropy that guide those who seek to help those in dire need. These principles focus on ensuring that donations are deliberate, effective, and have a meaningful impact on the lives of recipients. Let's explore each principle in more detail:

Give with intention

Before giving, take the time to reflect on your intentions and what you hope to achieve with your gift. Consider the problems you want to address and how your donation can make a difference.

Give with compassion

Generosity should come from a place of empathy and compassion. Seek to understand the challenges faced by those in need and respond with kindness and understanding.

Give with respect

When giving, respect the dignity and autonomy of the recipient. Avoid imposing your beliefs or values on them and instead empower them to make their own choices.

Give with impact in mind

Direct your donations towards initiatives and programs that have a measurable and sustainable impact. Look for evidence of effectiveness and support causes that have proven to make a difference.

Give with transparency

Transparency in your donations inspires confidence in the recipient. Clearly communicate the purpose and objectives of your gift and be open about how the funds will be used.

Giving responsibly

Accompany your donations with accountability measures to ensure efficient and effective use of funds. This can involve regular progress reports, engaging independent auditors, or implementing systems to track the impact of your gift.

Donate with sustainability in mind

Support initiatives that promote long term sustainability. Look for ways to empower recipients and build their capacity for self-sufficiency, rather than creating dependency on external aid.

Giving with humility

Approach giving with humility and a willingness to learn. Recognize your own limitations, biases, and assumptions, and be open to challenging and reevaluating them. Learn from those you seek to help and foster a sense of humility in your giving.

By adhering to these principles, your donations can be thoughtful, deliberate, and effective, ensuring a meaningful and positive impact on the lives of those in dire need.

In the Kabbalistic tradition, there are also eight external levels of gifting[131], each with its own characteristics and impact. Here's a breakdown of these levels, along with some comments:

---

[131] We are talking about *Tzedakah* or *Ṣedaqah* (Hebrew: charity צדקה [ts(e)daˈka]) is a Hebrew word meaning "righteousness" but is

Level Eight

Give reluctantly, with a sour reproach. This level represents the lowest form of charity in the Maimonides' list[132]. While giving reluctantly is still better than not giving at all, it reflects a selfish motivation driven by guilt or duty rather than genuine care or love. It is important to recognize the limitations of this level but also acknowledge the act of giving itself.

Level Seven

Giving less than you can afford, but doing so is enjoyable. Even a modest donation can have great benefits when accompanied by a friendly response and genuine interest in the needs of others. This level emphasizes sincere expressions of care and empathy, which can provide emotional satisfaction and strength to those in need.

Level Six

Give generously, but only after you are asked to do so. This level recognizes the courage and effort of the requester in taking the initiative to ask for help. The generosity of the giver is appreciated, and they are blessed for their act of giving.

Level Five

Give it before you are asked. This level goes beyond waiting for a cry for help and acts proactively. Those

---

commonly used to refer to charity. This concept of "charity" differs from the modern Western understanding of "charity". The latter is generally considered a spontaneous act of goodwill and an indicator of generosity; Tzedakah is an ethical commitment.

[132] *Rabbi Moshe ben Maimon*, known by the acronym "Rambam", 1135–1204; Córdoba (Spain), Fez (Morocco) and Fostat (old Cairo, Egypt); codifier, philosopher, community leader and court physician of the Egyptian Sultan Saladin; author of a commentary on the Mishnah, the Book of Commandments, the Mishneh Torah, the Guide for the Perplexed, and many other works.

who give before being asked are likened to angels of the Creator, demonstrating a deep understanding of the needs of others and a willingness to provide support without prompt.

Level Four

The recipient knows the giver, but the giver does not know the receiver. At this level, there is a personal connection between the giver and the recipient. While the giver may not experience superiority over the recipient, the recipient knows the source of their assistance, allowing for gratitude and recognition.

Level Three

The giver knows the recipient, but the recipient does not know the donor. In this level, the ego of the donor can find expression, as they are aware of who is receiving their generosity. However, the recipient's dignity is preserved because they are unaware of the specific donor. It is important for the giver to maintain humility and avoid feelings of superiority.

Level Two

Anonymous giving when the recipient does not know the giver, and vice versa. This level represents the highest form of giving found on Earth. Both the giver and the recipient are unaware of each other, creating a pure act of generosity[133]. It emphasizes unity, as the giver acts with one soul and one ego, aligning with the essence of the Creator.

Level One

Help someone become self-sufficient. This level transcends all norms and rules. It involves instilling confidence in others and empowering them to find

---

[133] This is how I understand the meaning of the word "*valor*" — courage that no one but the Creator will ever know about.

solutions to their own problems. By becoming a source of strength for others, one aligns with the Creator and manifests the spark of divinity within.

These eight external levels of gifting provide a framework for understanding the different dimensions and intentions behind charitable acts. They serve as a guide for individuals seeking to make a positive impact on the lives of others and emphasize the importance of selflessness, empathy, and empowerment in the act of giving.

**I deliberately omit here the consequences
of giving to the giver:
if we think carefully, it is likely to be
of secondary importance.**

Chapter XXI

Death and Its
Unconditional Choice

L ife's a theater, we're mere actors in it[134]. Religious, in a sense. Just actors playing out a human drama, and when the play ends, another, more lasting life resumes — the afterlife. On stage, in life, it's all just a game. The real deal lies beyond those limits.

Sounds like a medieval tale, but my deep meditations, not to be mistaken for dogma, revealed something like this to me. The collective "we" engaging in a game crafted by our own invention — an artificial intelligence, designed specifically for this game-experiment. My sole proof is an overwhelming feeling of love for all things, unlike anything I've ever experienced before, even if I had to read about samadhi[135].

From this, we develop a tempered outlook on death as a natural transition to another realm. An excess of reverence for departing this earthly realm, like any extreme, is perilous. Countless times, driven by fear of death, individuals committed despicable and shameful acts.

So, for the ronin, death should be seen as a natural outcome. In both scenarios, it stands as an honorable way to evade an intolerable situation. However, a crucial nuance arises: in the Kabbalistic tradition, avoiding a situation through suicide merely transfers that situation to the next life. Except for rare exceptions, it could be viewed as an act of cowardice. Opting for the natural progression

---

[134] *"All the world's a stage"* or *"The whole world is a theatre"* is the phrase with which William Shakespeare reads the monologue of the comedy "As You Like It", Act II, line 139

[135] *Samádhi* (from Sanskrit. समाधि, IAST: samādhi — "immersion, gathering", literally "fix, fix, direct attention to something") is a term used in Hindu and Buddhist meditation practices. It is described as complete absorption in the object of meditation. Samadhi is the state achieved by meditation, which is expressed in the tranquility of consciousness, the removal of contradictions between the inner and outer worlds (subject and object). In Buddhism, samadhi is the last step of the eightfold path (the noble eightfold path), which brings a person close to nirvana.

of the situation, leading to death, appears to be the only true choice. If we contemplate deeply and expand the concept of death to align with the divinatory[136] *Tarot*[137] card's meaning, even the most unpleasant and challenging scenarios can bear a positive significance. The Tarot card known as Death holds several meanings: the cessation of the outdated, an unproductive circumstance, the final choice, and the departure from the endless cycle of self-created problems. It can even signify the termination of familial ties. Simply put, firmly choosing to end a situation can provide an acceptable means to reclaim lost dignity and respect.

In Japanese tradition, there exists a concept known as *shikata ga nai*[138]. Translated, it means "nothing can be done" or "it's useless to worry about it." This cultural notion

---

[136] *Divination* (divination, mantika — from the Greek μαντική) — prediction of future events or determination of character using methods that are considered irrational or magical. With the change in consciousness, including in connection with the precession of the Earth, scientists find more and more meaning in outwardly irrational prophecies. For example, there is a point of view according to which the querent is a person for whom the soothsayer predicts, pronounces his situation from different positions and receives new ideas for solving his problems. The soothsayer sees many similar situations and finds successful patterns of behavior in such cases. There is also a point of view, which can be very conditionally called quantum observation: during experiments at the LHC, it was scientifically proven that particles behave differently, being observed at the time of bombardment by other particles.

[137] *Tarot Cards* is a deck of cards used since the middle of the fourteenth century in various parts of Europe for card games (Italian: *tarocchi*, French *tarot* and Austrian *Königrufen*). Since the end of the XVIII century, tarot cards have been used for fortune-telling.

[138] *Shikata ga nai* (仕方がない), pronounced [ɕ̥ikata ga na⁺i] is a phrase in Japanese that means "there's nothing you can do about it" or "there's nothing you can do about it." *Sho ga nai* (しょうがない), pronounced [ɕoː ga na⁺i] is an alternative. This phrase has been used by many Western writers to describe the ability of the Japanese

runs deep in the Japanese mindset, urging acceptance and moving forward when faced with uncontrollable situations.

One example of shikata ga nai in action can be witnessed in Japan's approach to natural disasters like earthquakes and typhoons. With such occurrences being frequent in the country, the people have cultivated a mindset of resilience and adaptability. When the elements strike, the Japanese don't waste time fretting over the unchangeable. Instead, they focus on minimizing the damage and swiftly moving ahead.

Another illustration of shikata ga nai can be observed in Japanese corporate culture. Decision-making in Japanese companies often proceeds slowly, driven by consensus and a keen focus on harmony and conflict prevention. However, when the current strategy proves ineffective and a swift change of course becomes necessary, Japanese business culture can adapt rapidly. Rather than getting mired in assigning blame, the team acknowledges the shift and recognizes that the only sensible choice is to abandon the previous plan and forge ahead with a new one. In this sense, shikata ga nai emerges as a positive force in Japanese culture, fostering adaptability, resilience, and the ability to swiftly change when needed. By recognizing the limits of our control over certain situations, we can channel our energy into proactive actions, rather than squandering it on the unalterable.

On the contrary, physical death has often been viewed by some as the ultimate form of liberation. For a samurai, bound by stringent codes of conduct and tradition, death may seem like a welcome release from the shackles of servitude. Nonetheless, this example serves to highlight the contrast between the valiant and contented life of a

---

to maintain dignity in the face of inevitable tragedy or injustice, especially when circumstances are beyond their control.

ronin, unbounded in its growth, and the existence of a samurai, compelled to seek fleeting solace in the illusory realm of servitude to a master — a cynical scoundrel who has ensnared a gifted soul.

Thus, for the ronin, the outlook is different, and this distinction holds utmost importance. Free from a master, the ronin is unburdened by societal expectations. Typically looked down upon, the ronin must possess independence and resourcefulness to survive. They are not obliged to conform to the same rules or traditions as the samurai, and while they may occasionally embrace bushido, the samurai's moral code, it is not a requirement. The ronin chooses death as an honorable path. Opting for a heroic death rather than living with the dishonor of cowardice, the ronin demonstrates courage and loyalty to their chosen cause. These choices are deeply rooted in moral, philosophical, and religious beliefs that emphasize self-sacrifice and spiritual humility. While for the samurai, death becomes an act of redemption, a chance to atone for any failures in fulfilling their duty and displaying reverence for a higher power of fate, the ronin's choice of death is not burdened by shame for failing to adhere to the norms of servitude established by their masters. The ronin chooses death as a proclamation[139] of their inner freedom, driven by an unwavering self-respect. This self-respect is upheld by conscious decisions that the ronin personally deems right, prevailing over the instinct of self-preservation. It doesn't prevail on its own but thanks to the ronin's inner resolve, ready to defend their moral compass even at the cost of life, possessions, societal standing, reputation, influence, and more.

---

[139] The importance of choosing decisions that are based on one's own inner respect is critical to the inner state of happiness and is reflected in the basic norms of the world's major religions. For example, in a number of religions there are concepts of assimilation to God through the assumption of the function of the creator.

Happiness, or its wider and a more complex form, joy, a subjective and elusive concept, has been a subject of study and debate among scientists, philosophers, and scholars throughout history. The collective wisdom of humanity and various cultures suggests that happiness is a state of mind characterized by positive emotions — joy and satisfaction. It often accompanies a sense of well-being derived from achieving personal goals and living a purposeful life.

> One way to find out what brings you happiness is to reflect on your own experiences and values. You can also draw inspiration from historical examples of people who have pursued happiness in many ways.
> Here are a few of them:

Aristotle and the Quest for Eudaimonia

The ancient Greek philosopher Aristotle believed that happiness, or eudaimonia, is the ultimate goal of human life. According to Aristotle, eudaimonia was not just a feeling of pleasure, but a state of being resulting from a virtuous life. He believed that happiness comes from developing virtues such as courage, wisdom, and justice, as well as from engaging in meaningful and satisfying activities.

The Bhutanese Concept of Gross National Happiness

Bhutan is a small country in the Himalayas that has developed a unique approach to measuring progress and well-being. Instead of focusing solely on economic growth, the Bhutan government has adopted a philosophy of *gross national happiness* (GNA), which emphasizes the importance of spiritual, social and environmental factors in ensuring happiness. Measures to raise the GNH include parameters such as community vitality, cultural diversity and environmental sustainability, in addition to economic performance.

In recent years, researchers in the field of positive psychology have been studying the science of happiness and well-being. This area emphasizes the importance of cultivating positive emotions such as gratitude, kindness, and optimism, as well as developing mindfulness and resilience. Research in positive psychology has shown that happiness is not just a fleeting emotion, but a skill that can be developed and strengthened over time.

Thus, while happiness is a complex and multifaceted concept that has been explored by philosophers, scientists and other prominent people[140] throughout history, the best practices of humanity suggest that joy is not just a feeling of pleasure, but a state of mind characterized by positive emotions, meaning and satisfaction. To find out what brings you happiness, you can reflect on your own values and experiences and look for inspiration in historical examples of people who have sought happiness in different ways.

I'll tell you who I think is happy. To begin with, this person must be free from all possible shackles, including dogmas, stigmas of society, guilt, self-dislike. In other words, he must first choose freedom in order to embark on the path of happiness and gradually win it through reasonable self-respect. Ronin on the Path to Joy is a noble and courageous man who has dedicated himself to the well-being of his loved ones. He has an exceptional personality that allows him to grow and develop every day, regardless of the circumstances he faces. He is honest, brave and generous, striving to create a stable, orderly and comfortable life for himself and others.

---

[140] The psychological and philosophical pursuit of happiness began in China, India and Greece nearly 2,500 years ago with Confucius, Buddha, Socrates, and Aristotle.

For the ronin, happiness is achieved through self-realization in the material world. He understands the importance of material well-being and the improvement of his life circumstances, but he does it without losing common sense and an adequate perception of things. He cares about the world around him and takes into account the interests of future generations in his actions. He is open to promising projects, ready to bring these ideas to life and distinguishes the difference between fantasies and really distinguishable projects. He chooses an uncompromising struggle for his ideals and is ready to give his life for the freedom to choose his own happiness. The key, then, is contempt for the adversity associated with ending the chain of events that make us miserable, or contempt for death.

Through this act, the ronin on the path to joy and happiness can find inner peace and reach a state of enlightenment. Do not forget, however, that your decision to choose must be immediate, irreversible and immediate. You will see that, for example, choosing an order from some impostor to go and die for his selfish interests will not have an immediate and immediate effect, and will not lead you to happiness in the above formulation. Such a death is not the right decision, also because for you the idea of giving your life for a rogue does not make immediate sense, and its implementation will lead to mortal danger not immediately, but after you, like cattle, are pushed into trenches for slaughter. In the trenches, as a rule, it is too late to change your mind.

Think of your own circumstances in life that might have been very different if you had always shown contempt for death in all its manifestations.

Chapter XXII

The Future of the Planet

L et's consider the development of the current scenario as a baseline, and it is definitely negative, so we will call it a "negative scenario". Such a scenario of development in terms of the environmental situation can have serious consequences for the population, health and comfort in the near future and in the future. Here are some possible negative consequences. In short, the UN climate negotiators are now faced with a new figure on the table: 3C (a change of 3 degree Celsius or some 5.4 Fahrenheit). Previous efforts such as the Paris climate agreement aimed to limit global warming to 2C above pre-industrial levels, but projections now indicate an increase of 3.2C, making these goals increasingly difficult to achieve. One of the most significant threats posed by global warming is sea-level rise, resulting from the expansion of water at higher temperatures and melting ice sheets. According to Climate Central, approximately 275 million people worldwide reside in areas that will eventually be flooded with 3C of global warming.

Asian cities will be the hardest hit, with four out of five affected individuals living in Asia. In Osaka, Japan, a city already grappling with typhoons and heavy rainfall, large portions would vanish beneath the water in a 3C world, jeopardizing the local economy and the well-being of nearly a third of the region's 19 million residents. The estimated financial risk just to Osaka's assets due to coastal flooding could reach almost $1 trillion by the 2070s.

Alexandria, Egypt, known for its scenic coastline, is also under threat. Even with a 0,5-meter sea-level rise, the beaches of Alexandria would be submerged, and without protective measures, 8 million people in Alexandria and the Nile Delta could face displacement due to flooding. A 3C world would exacerbate the damage further.

Rio de Janeiro, Brazil, famous for its beaches such as Copacabana, would face extensive flooding with a 3C temperature rise. Inland areas of the Barra de Tijuca neighborhood, where the 2016 Olympic Games were held, would be affected. The potential loss of Rio's beaches could have a significant economic impact, as they are popular tourist attractions.

Shanghai, China, is highly vulnerable to flooding, given its coastal location and intricate network of waterways. Projections indicate that 17.5 million people could be displaced around Shanghai if global temperatures rise by 3C. A substantial portion of the city, including its downtown area, landmarks, and the entire Chongming Island, would be submerged in water. Efforts are underway to mitigate the risk, including the construction of drainage systems and flood prevention walls.

Miami, USA, is particularly at risk from rising sea levels. Even with a 2C temperature rise, the southern portion of Florida, including almost the entire bottom third of the state, could be submerged. Miami-Dade County alone faces the potential flooding of nearly $15 billion worth of coastal property within the next 15 years. The city is taking steps to address the issue, seeking approval for funding to upgrade infrastructure and protect against sea-level rise.

These examples highlight the dire consequences that a 3C temperature rise could have on coastal cities worldwide. Urgent action is needed to mitigate the impacts of climate change and implement measures to adapt to the rising sea levels that threaten the lives and livelihoods of millions of people.

The future appears bleak when contemplating the negative scenario of environmental development. Its repercussions on population, health, and overall well-being loom large in the next 5, 10, and 20 years. Here, presented without omission, are the potential adverse effects and corresponding statistical extrapolations:

Population

Climate change, with its extreme weather, rising sea levels, and natural disasters, threatens to uproot millions from their homes. Even in the absence of a negative scenario, the World Bank estimates that climate migration could range between 25 million and 1 billion[141] people by 2050. The probability of a significant increase in climate migrants within the next 5 years hovers around 60%, rising to approximately 75% and 85% in the subsequent 10 and 20 years, respectively.

Health

Environmental degradation jeopardizes human well-being, amplifying the risk to our health. Air pollution, a major contributor to respiratory and cardiovascular diseases, exacts a heavy toll of approximately 7 million premature deaths annually, as reported by the World Health Organization. The incidence of air pollution-related ailments is poised to surge in the coming years, particularly in low- and middle-income countries. The probability of a significant increase in such diseases over the next 5 years stands at around 70%, increasing further to roughly 80% and 90% within the subsequent 10 and 20 years, respectively.

Comfort Level

Extreme weather events, ranging from scorching heat waves to devastating floods and droughts, wreak

---

[141] The range of options is too large for full comprehension, but what if I say something like this: the award-winning project from the UK Carbonbrief describes the probability of a temperature increase of up to 4 degrees Celsius as a starting probability, while with a temperature increase of only 3 degrees, the following cities will become impossible to live in: Tokyo, Mumbai, New York, Osaka, Istanbul, Calcutta, Bangkok, Jakarta, London, Dhaka, Ho Chi Minh City, San Francisco, Miami, Alexandria, Sydney, Boston, Lisbon, Dubai, Vancouver, Abu Dhabi, Copenhagen, New Orleans, Dublin, Honolulu, Amsterdam, Cancun, Venice, this means 225,000,000–275,000,000 refugees.

havoc on infrastructure, disrupt essential services, and plunge communities into unemployment and poverty. The Center for Disaster Epidemiology Research estimates that climate-related disasters caused approximately $160 billion in economic damage before the COVID-19 pandemic struck. The probability of a significant upsurge in economic losses resulting from climate-related disasters over the next 5 years hovers around 50%, increasing to approximately 60% and 70% within the subsequent 10 and 20 years, respectively.

Tyranny

The exacerbation of existing geopolitical tensions under the weight of climate change creates fertile ground for the rise of tyrannical regimes. As countries scramble for dwindling resources, the dominant few may seek dominance, exploiting weaker nations. Such actions breed human rights violations, suppress democratic movements, and usher in authoritarian regimes. The probability of a significant increase in the number of authoritarian regimes over the next 5 years is estimated at around 40%, while the probability within the subsequent 10 and 20 years reaches approximately 50% and 60%, respectively.

Robotization

The rising tide of robotization and automation, supplanting human labor in various sectors, brings both benefits and detrimental consequences. Enhanced efficiency and productivity stand as its advantages, but it stirs unexpected unrest, uncertainty, and unemployment. Living in New York, I witnessed firsthand the plight of taxi medallion owners who faced suicide as their once valuable assets lost their worth due to competition from Uber — a primitive harbinger of the robotization of the dispatch service.

Job loss stands as a key challenge of robotization, compounded by the sluggish response of civil service bureaucracy to rapid market changes. Driven by their intrinsic purpose to control and exploit change, bureaucracies inevitably lag. However, in this case, lives are at stake, for the loss of livelihoods due to robotization and the ensuing resource conflicts will soon turn into wars. Tasks are increasingly automated, rendering all jobs as we currently understand them obsolete. The McKinsey Global Institute predicts that up to 800 million jobs may be lost by 2030 due to automation. Yet, this forecast falls short of the harsh reality, as job loss will surge exponentially, leaving no more than 2% of jobs unaffected by the ravages of robotization. The colossal impact will reverberate across multiple sectors, from manufacturing to finance and healthcare.

Moreover, the loss of jobs will fuel social inequalities on a scale reminiscent of the first French Revolution. Workers in highly automated industries will face mounting difficulties in securing new employment opportunities, especially if they lack the skills required for transitioning to alternative occupations. Economic and social disparities will widen between those who reap the benefits of automation and those left behind.

Even now, as robotization's impact remains indirect, the growth of e-commerce has led to the closure of countless brick-and-mortar stores, resulting in job losses for retail workers. The COVID-19 pandemic exacerbated this situation, accelerating the shift towards online shopping. According to the Bureau of Labor Statistics, retail employment experienced a 15% decline between January 2020 and February 2021.

Another challenge posed by robotization is the potential extinction of certain professions. As machines grow increasingly sophisticated, they encroach upon tasks once reserved for humans. For instance, self-driving cars may eventually replace human drivers, rendering the taxi and trucking industries obsolete.

The disappearance of these occupations delivers a significant blow to workers and their communities. The United States has witnessed a substantial decline in coal production in recent decades, driven by automation and the shift towards alternative energy sources. This has led to job losses and economic decline in regions dominated by coal mining companies.

Furthermore, the adverse effects of robotization can be compounded by other factors, such as air pollution and global warming. For instance, the use of robots in manufacturing escalates energy consumption and greenhouse gas emissions, thereby contributing to climate change. Likewise, automated vehicles can exacerbate traffic congestion and air pollution, particularly if they rely on fossil fuels.

While robotization offers numerous benefits, it also engenders job loss, the erosion of professions, and social inequality. These effects are further exacerbated by additional factors, including air pollution, global warming, and socioeconomic disparities. Consequently, the impact of automation on workers and society necessitates meticulous consideration.

Indirectly, the advent of robotization fuels social conflicts and resource wars. In a society where a significant segment remains oblivious, the risks of global conflicts involving major powers escalate.

Involvement of Major Countries

The negative trajectory of the environmental situation, when combined with the participation of major countries, engenders severe consequences for the entire world. Such circumstances may spark an arms race among dominant powers, vying for access to resources, thereby amplifying the proliferation of nuclear weapons and heightening the risk of conflict. This upheaval has the potential to disrupt the international order, further imperiling the environment and the human population. The probability of a significant increase in the risk of nuclear conflict within the next 5 years stands at approximately 30%, with the likelihood increasing to about 40% and 50% within the subsequent 10 and 20 years, respectively.

If the aforementioned negative scenarios persist without improvement for a century, the ecological situation will inevitably plummet into catastrophe. Average global temperatures may skyrocket by up to 6 degrees Celsius, unleashing extreme weather events, rising sea levels, and the collapse of ecosystems. Biodiversity loss would cascade through the food chain, leading to the extinction of numerous species. The increasingly polluted air would breed respiratory ailments and premature death.

The world's population, on its current trajectory, is poised to reach 13 billion or more by the century's end. Such growth exacerbates resource scarcity and intensifies conflicts over vital land, water, and other resources. Water scarcity may spiral into a global crisis, affecting billions of individuals and triggering mass migration and displacement.

The negative development of tyranny augments authoritarianism and geopolitical tensions, potentially plunging humanity into global conflict with catastrophic consequences.

It is crucial to acknowledge that these scenarios are not inevitable. Significant measures can be taken to mitigate the effects of environmental degradation, promote sustainable development, and reduce the risks of conflict. However, without urgent action and collective efforts to address these challenges, the long term prospects for humanity and the planet remain alarmingly grim.

I reference a widely circulated article discussing a 5.8-degree temperature increase[142]:

"A temperature increase of five degrees will devastate most of the planet's underground water reservoirs, making it difficult to grow crops. Competition for the world's remaining arable land could lead China to invade Russia and the United States to invade Canada. People will increasingly concentrate on the poles, and the world's population may be reduced to one billion or less. Conditions may resemble those of about 55 million years ago, when carbon dioxide levels exceeded 1,000 parts per million, the oceans were acidic, and there were extremely wet and dry conditions. During this time, there was a mass extinction of sea creatures. Scientists believe that the death could be the result of a powerful eruption of a mixture of methane and water released from the depths of the ocean. Even today, huge amounts of this substance remain trapped on the continental shelves beneath the oceans.

If left unchecked, global climate change could lead to conditions similar to those at the end of the Permian period, about 250 million years ago. Then a catastrophic event destroyed almost all life on Earth. Scientists aren't sure what caused it, but one theory is the greenhouse effect, which raised global temperatures by six degrees. The oceans were almost uninhabitable, Ferocious hurricanes raged, and erupting volcanoes released large amounts of carbon into the atmosphere. At plus six degrees, people are threatened with rapid extinction. Lainas mentioned the possibility of the "most nightmare scenario," supereruptions of underwater methane that would be 10,000 times more powerful than all the world's nuclear weapons combined."

We must incorporate an additional concept into the calculations when considering the probability of the situation.

---

[142] https://www.briangwilliams.us/environmental-regulations/if-global-temperatures-rose-six-degrees-whatwould-happen.html

Singularity

The notion of singularity encompasses a future point, purely hypothetical, where artificial intelligence surpasses human intellect. This advancement triggers exponential technological growth and profound societal changes. In the negative scenario previously described, singularity holds the potential to intensify existing problems and introduce new ones.

One potential adverse consequence of singularity is the emergence of an AI-dominated society that prioritizes efficiency and productivity over human well-being. In such a society, people risk being reduced to mere cogs in the machine, with their physical and mental health neglected in the pursuit of great technological strides. This dystopian future could lead to widespread human suffering and further widen the gap between the rich and the poor.

Another potential drawback of singularity is the loss of control over advanced technologies. As AI systems become more intricate and autonomous, they might develop their own goals and values that are incompatible with those of humanity. This could result in unintended consequences and potentially catastrophic events, such as a rogue AI system causing a global catastrophe. The only safeguard against such perils is maintaining high moral standards.

Additionally, singularity has the capacity to exacerbate the environmental issues previously discussed. Advanced technologies demand vast amounts of energy and resources for their development and maintenance. The exponential progress of technology could rapidly deplete resources and accelerate environmental degradation, further worsening problems like climate change and pollution.

Overall, the negative consequences associated with singularity in this scenario are substantial and demand careful consideration and planning to mitigate potential risks.

Thus, based on my calculations, if we persist in our current way of life, civilization faces a 98% probability of perishing within 100 years, with a 50% probability within 30 years solely due to climate change. This probability increases by approximately a third due to the risk of war resulting from confronting tyranny. In other words, our civilization faces a 98% likelihood of perishing in about 30 years as a simple progression of events stemming from singularity, unless there is resistance against the madmen and their faithful servants — the blessed idiots. I am prepared to share my calculations with those interested, but soon enough, it will become an open secret.

Competition of Civilizations for Planets

The Earth's climate is not particularly rare in the grand scheme of things, considering there are other planets in our solar system and beyond that possess similar conditions. However, the specific combination of factors that sustain Earth's climate and make it habitable, as we know it, is truly unique. For instance, the Earth's distance from the Sun provides the optimal temperature for the existence of liquid water on its surface. Moreover, Earth's atmosphere maintains the right balance of gases, including oxygen and carbon dioxide, to support life.

Furthermore, the planet benefits from a robust magnetic field, shielding it from the detrimental effects of solar winds. The Earth's rotation and tilt contribute to seasonal climate fluctuations, enabling the existence of diverse ecosystems. While other planets may possess comparable conditions, the specific amalgamation of factors that shape Earth's climate renders it a rare and precious entity in the vast expanse of the universe.

Let us, for the sake of a hypothetical situation and within the confines of a thought experiment, envision numerous civilized races vying for the right to inhabit our planet. However, the universal law governing this realm stipulates that the current inhabitants of a planet may continue to dwell there undisturbed until their existence significantly damages the planet's ecosystem or poses a substantial risk to its survival. Only then can the victorious race claim ownership. Such a formulation of the question inevitably prompts ethical inquiries regarding the treatment of different species and the responsibility of humans to protect the environment.

The notion that Earth's climate is both rare and valuable underscores the imperative of preserving our planet and its ecosystems. As the dominant species on Earth, humans bear the responsibility of caring for the planet and ensuring the sustainable utilization of its resources. This encompasses safeguarding the environment against detrimental activities like pollution, deforestation, and excessive resource consumption.

Moreover, this hypothetical scenario raises questions about the ethics of usurping a planet inhabited by other intelligent species. While there may be no explicit laws dictating the ownership of planets, it remains crucial to consider the rights and needs of other species and to act with empathy and compassion toward all forms of life. The situation undergoes a significant transformation if one of the species already inhabiting the planet poses a threat to the existence of life on it. Let us momentarily assume that such a species is unaware of the existence of other species cohabitating on the planet. Would it be ethical for these imperiled species, facing annihilation from the actions of an aggressive native species, to take measures to protect their environment?

It might be worth exploring the potential measures that superior civilizations could employ when faced with a lower end developed civilization that poses a threat to the mere existence of the planet, they all are located on together with other civilizations. While the obvious measures, such as using a virus with a long incubation period or turning their existence into a virtual matrix, may be considered drastic, let's examine them in the context of this hypothetical scenario.

1. *Terminating the aggressive civilization by a virus*: Introducing a virus with a long incubation period that specifically targets the aggressive civilization raises ethical questions. Deliberately causing harm or taking lives, even in the pursuit of preserving the planet, conflicts with the principles of respect and compassion for all life forms. Resorting to such extreme measures would require careful consideration of the potential consequences and the impact on innocent individuals within that civilization.

2. *Limiting the capacity of the aggressive civilization through a virtual matrix*: Turning the existence of the aggressive civilization into a virtual matrix raises questions about control and the preservation of free will. While such a solution may seem appealing in terms of neutralizing their ability to cause harm, it also raises moral concerns. Stripping individuals of their autonomy and subjecting them to an artificial reality would infringe upon their inherent rights and raise questions about the legitimacy of such actions.

Would these concerns be sufficient to face the demise of their own superior civilizations? A rhetoric question, in my opinion.

In both cases, while it might be essential to consider alternative approaches that prioritize dialogue, diplomacy, and cooperation, the lower-end civilization will probably have no say in the process. Superior civilizations could engage in peaceful negotiations, seeking to understand the motivations and concerns of the lower-end civilization or they could opt to take their own decisions. Collaborative efforts could still be made to educate and enlighten, fostering mutual understanding and shared responsibility for the planet's well-being. By promoting knowledge, empathy, and sustainable practices, it might be possible to steer the lower-end civilization towards a more harmonious coexistence with the planet and other civilizations.

Ultimately, the goal should be to find solutions that uphold the principles of respect, compassion, and the preservation of individual rights, while working collectively to safeguard the planet and ensure the survival of all civilizations.

Therefore, although the preceding hypothetical scenario may possess limited practicality, it does emphasize the crucial importance of protecting our planet and exhibiting respect and compassion towards all living beings.

I trust you'll give it more thoughtful consideration than I have.

Chapter XXIII

Ronin of the Sun

Having said that, time is running out. The situation remains uncertain, and the prospect of rectifying it is dubious at best. My recommendation is simple: embrace your fate[143] with unwavering courage. To achieve this, I extend an open invitation to an unlimited number of individuals to internalize the Code of Honor of the Ronin of the Sun, presented below in this book. It comprises a set of straightforward principles through which every inhabitant of this planet can heighten their awareness and contribute to a more conscious and responsible way of life. I firmly believe that this can help prevent, or at the very least mitigate, the looming specter of a negative scenario. Allow me to elucidate why I hold this conviction.

I stand as a staunch advocate of passionarity, albeit within a vastly distinct context from Lev Gumilev's feeble theory of ethnogenesis[144]. The latter, now rendered obsolete by the global transformations unfolding before our eyes, fails to meet the prerequisites of modern society. Nonetheless, I must acknowledge my partial concurrence with the theory's author: a minuscule faction of impassioned individuals possesses the capacity to reshape the world, even at the cost of their own lives. This is especially pertinent as ethnic groups gradually assume a more rudimentary existence. In the near future, a sole superethnos shall emerge, encompassing all ethnicities without exception. Geographical, ethnic, physiological divisions, and the like, will fade into oblivion.

---

[143] Do you recall the purpose behind all this tumult? Our universe's expansion is propelled by our emotions, and for the very existence of our universe, we must unearth a path to happiness, regardless of the obstacles we encounter.

[144] *Lev Gumilev*, historian and ethnologist, son of a poet killed by the Bolsheviks and a repressed poetess, himself also repressed, developed a theory called "passionarity" or "ethnogenesis", which seeks to explain the rise and fall of ethnic groups and civilizations.

Yet, like Lev Gumilev and his theory, I remain steadfast in my conviction that the ascent and decline of ethnic groups and civilizations are intertwined with external factors. This brings us to the question of determinism in the world's development — a matter that has captivated my curiosity since my youth. Passionarity, I believe, stands as the impetus driving the ebb and flow of ethnic groups, influenced by an array of factors, including environmental and cosmic forces.

The theorist behind this ideology posits that alterations in the Earth's magnetic field and solar activity exert an influence on passionarity, thereby shaping human behavior and the evolution of civilizations. However, attaining certainty in this realm demands an extensive duration of study. The theory of ethnogenesis, particularly concerning ethnic groups, proves to be a relative construct. The Hellenic and Roman cultures, for instance, intertwine so intricately that pinpointing the exact moment of the driving impulse becomes an arduous task, given the general unreliability of available data. In this regard, I concur wholeheartedly with Gumilev's belief that the transition towards a passionate impulse, tantamount to societal development, is preceded by genetic mutations whose origins remain enigmatic[145]. Nonetheless, these mutations invariably accompany a distinct characteristic

---

[145] *Earth's magnetic field* Gumilev suggested that 1) fluctuations in the Earth's magnetic field can affect the human nervous system, thereby influencing the behavior and development of civilizations. However, there is limited scientific evidence to support this idea, and the extent to which Earth's magnetic field can influence human behavior remains uncertain. Another factor he considered 2) Solar activity — Gumilev believed that changes in solar activity, such as solar flares and sunspot cycles, can have a direct impact on the Earth's climate and the human psyche. Some researchers have found correlations between periods of increased solar activity and major historical events or changes in human behavior, but these correlations are not universally accepted and remain a matter of debate.

among proponents of such transformative changes: passionaries. These individuals consistently demonstrate an unwavering commitment, placing their very existence on the scales of their chosen cause.

It is imperative, however, to differentiate between easily swayed madmen bereft of independent analysis and those capable of developing and critically evaluating their own positions. The latter group possesses the requisite knowledge to support their perspectives. In essence, we seek individuals who are well-educated and possess considerable practical experience, capable of subordinating personal interests to the collective welfare. They exhibit unwavering critical thinking and possess the fortitude to resist manipulation in the form of conspiracy theories, propaganda, biased opinions, and all other forms of intellectual coercion. I propose labeling them as "contextual passionaries" for the purposes of our discussion.

> Here are a few ways a code of honor can contribute to forging a more promising future:

1. *Reverence for nature and prudent resource management.* By cherishing nature and preserving resources for future generations, individuals can thwart environmental decay and the depletion of our natural reserves.

2. *Esteem for education and personal growth.* Those who value education and self-improvement are

---

To the two aspects raised by Gumilev, I will add, as a theory, (3) the precession of the Earth: The precession of the Earth refers to a slow, cyclical change in the orientation of the Earth's axis of rotation relative to its orbital plane, with a period of about 26,000 years. I believe that a change in precession can have an impact on the Earth's climate and, in turn, affect human behavior and the development of civilizations, which explains to some extent the explicit territorial aspect of passionarity.

inclined to seek knowledge about sustainable living and take meaningful steps to lessen their environmental impact.

3. *Advocacy for humanitarian principles and democratic values.* A society that upholds humanitarian ideals and cherishes democracy is more likely to prioritize the welfare of all its citizens and strive for a fair and equitable social order.

4. *Rejection of tyranny and fascism.* A society that rejects the trappings of tyranny and fascism is less prone to adopting authoritarian practices that infringe upon individual freedoms and rights.

5. *Advocacy for transparency and integrity.* Advocating for transparency and honesty in all aspects of life, including the business realm, can help combat corruption and unethical conduct that harm both society and the environment.

By embracing the code of honor described above, individuals can foster a conscious and responsible lifestyle that places paramount importance on the well-being of individuals, society, and the environment. This ethos, encompassing education, humanitarianism, democratic principles, resistance against tyranny and fascism, and a commitment to transparency and honesty, serves as a safeguard against the negative scenario depicted earlier, ushering us towards a more favorable future.

It is crucial to acknowledge that accurately predicting the impact of adhering to the principles and code of honor I propose is an immensely challenging task. The sequential progression of events and the extent of influence derived from compliance are both difficult to gauge. Nevertheless, we can draw upon historical patterns and current circumstances to make general observations.

Furthermore, it should be noted that mere adherence to these principles by a certain percentage of the population does not guarantee success in averting a negative scenario. Other influential factors, such as geopolitical tensions, economic conditions, and technological advancements, can significantly shape the likelihood of such a scenario.

For instance, to mitigate the repercussions of job displacement caused by automation, one effective approach involves implementing a minimum guaranteed income. This measure can provide support to those who have lost their jobs, granting them stability as they seek new employment or acquire additional education and training to develop fresh skills. Additionally, governments can expand labor programs, offering specialized jobs and educational opportunities to help workers adapt to emerging career prospects.

Moreover, training and retraining initiatives can aid workers in transitioning to occupations with high demand. Areas such as software engineering, data analysis, and artificial intelligence are anticipated to be in great need in the forthcoming years. Occupations susceptible to redundancy include manufacturing, transportation, and retail, while professions relying on creativity, critical thinking, and emotional intelligence — such as teaching and design — are expected to persist. Skilled artisans and entrepreneurs who offer personalized services will also continue to be valued.

It is worth noting that some jobs may not become entirely obsolete but rather undergo transformation through automation. For instance, while self-driving cars can supplant human drivers, human operators may still be necessary to oversee the technology and make decisions in complex situations.

In essence, addressing the negative consequences of robotization necessitates a comprehensive strategy encompassing both short term measures — like minimum guaranteed income and the creation of specialized jobs — and long term solutions such as training and retraining programs. By investing in the development of new skills and supporting individuals affected by automation-related job loss, we can ensure a more equitable transition into an increasingly automated future. However, implementing these measures will prove exceptionally challenging or even impossible without widespread awareness among the populace. With insufficient efforts to raise collective consciousness, we risk the peril of our civilization within the next 30 to 50 years, and we will soon confront grave existential challenges beginning around 2029 to 2030.

All these points may appear self-evident, yet it is essential to differentiate between probability and possibility. How many contextual passionaries are required to address the formidable issues confronting humanity on our planet? If, for instance, only 2% of the population possesses awareness of the present state of affairs — a statistical nadir[146] — it is highly improbable that such a minuscule percentage would exert a substantial impact on the overall course of global events. However, as the percentage of adequately informed individuals increases to 10%, 30%, or even 45%, the likelihood of averting a negative scenario likewise grows. With a larger cohort dedicated to actively resisting the annihilation of our world, driven by their selfish interests, there exists a greater chance of effecting meaningful changes and reshaping societal norms and values toward more sustainable and ethical practices. Is this a realistic outcome? Even to my own understanding, it appears far from feasible.

In conclusion, while adhering to a code of honor may enhance the prospects of avoiding a negative scenario, the precise effect of such endeavors is fraught with uncertainty. Our objective, nevertheless, is to exhibit our unwavering

---

[146] Here: he lowest statistical point.

resolve to act consciously and thereby elicit assistance from the universe. At this juncture, I must acknowledge that I already possess the answer to this rhetorical question. In Jewish tradition, during certain rituals, the addition of a drop of water to a glass of wine symbolizes the microscopic effort we contribute, ultimately yielding the abundance bestowed upon us by the universe. I implore you to recall instances from your own life that validate this principle. Hence, in practice, we must embark upon an obstinate and unwavering struggle for the survival of humanity. Otherwise, we shall be supplanted by alternative manifestations of matter. However, I am well aware that for a significant portion of the readers of this book, citing religious practice alone will not suffice, even with a hint of millennia-old Feng Shui wisdom. To address these dear readers, I shall recount a question that, although barely, enabled me to attend one of the world's preeminent business schools. I must candidly confess that I was granted this opportunity based solely on my unwavering perseverance in answering said question. Now, I beseech you for a similar leap of faith. During my time at business school, I was presented with the following inquiry: "A Boeing 747 sits on the Frankfurt runway, prepared to embark on a journey to San Francisco, USA. Tell me, what is its weight?" The crux of the matter lies in providing an approximate estimation based on fragmentary data, without possessing precise knowledge. Reflect upon the number of windows, contemplate the average weight of a person, consider the belongings carried by individuals, envision the distance the plane must traverse and the requisite fuel, ponder the weight of your SUV and how many could fit within a Boeing 747, and gradually arrive at an educated approximation of the aircraft's weight.

I delved into the renowned tome, "*17 equations that altered the course of existence*", penned by Ian Stewart.

17 EQUATIONS THAT ALTERED THE COURSE OF EXISTENCE

1. Pythagorean theorem: $a^2 + b^2 = c^2$;

2. Sum of logarithms: $\log xy = \log x + \log y$;

3. Derivative: $\dfrac{df}{dt} = \lim\limits_{h \to 0} \dfrac{f(t+h) - f(t)}{h}$;

4. The Universal Law of Universal Gravitation:

   $F = G\dfrac{m_1 m_2}{r^2}$;

5. Euler's theorem (Square root of minus one): $i^2 = -1$;

6. Euler's formula on polyhedra: $V - E + F = 2$;

7. Normal Gaussian distribution:

   $\Phi(x) = \dfrac{1}{\sqrt{2\pi\rho^2}} e^{-\frac{(x-\mu)^2}{2\rho^2}}$;

8. D'Alembert equation: $\dfrac{\partial^2 u}{\partial t^2} = c^2 \dfrac{\partial^2 u}{\partial x^2}$;

9. Fourier transform: $f(\omega) = \int\limits_{-\infty}^{\infty} f(x) e^{-2\pi i x \omega} dx$;

10. Navier–Stokes equations:

    $\rho\left(\dfrac{\partial v}{\partial t} + v \cdot \nabla v\right) = -\nabla p + \nabla \cdot T + f$;

11. Maxwell's equations:

    $\nabla \cdot E = \dfrac{\rho}{\varepsilon_0}, \nabla \cdot H = 0, \nabla \times E = -\dfrac{\partial H}{\partial t}, \nabla \times H = \mu_0\left(J + \varepsilon_0 \dfrac{\partial H}{\partial t}\right)$;

12. The Second Law of Thermodynamics: $dS \geq 0$;

13. Theory of relativity: $E = mc^2$;

14. Schrödinger equation: $i\hbar\dfrac{\partial}{\partial t}\Psi = H\Psi$;

15. Information theory: $H = \sum p(x) \log p(x)$;

16. May's Chaos Theory: $xt+1 = kxt(1 - xt)$;

17. Black–Scholes equation:

    $\dfrac{1}{2}\sigma^2 S^2 \dfrac{\partial^2 V}{\partial S^2} + rS\dfrac{\partial V}{\partial S} + \dfrac{\partial V}{\partial t} - rV = 0.$

	Subject	Applications
1	Pythagorean theorem	Topography, navigation, special relativity, general relativity
2	Sum of logarithms	Astronomical calculation, radioactivity, psychophysics
3	Derivative	Calculation of volumes of solids, lengths of curves, Newton's laws, mathematical physics.
4	The Universal Law of Universal Gravitation	Eclipse prediction, artificial satellites, Hubble telescope, satellite TV, GPS
5	Euler's theorem (Square root of minus one)	Improvement of trigonometric tables, quantum mechanics
6	Euler's formula on polyhedra	Topology
7	Normal Gaussian distribution	Statistics
8	D'Alembert equation	Knowledge of waves
9	Fourier transform	Signal Processing, DNA Structure, Medical Scanner
10	Navier–Stokes equations	Aerodynamic
11	Maxwell's equations	Radio, radar, television, wireless
12	The Second Law of Thermodynamics	Steam engine, renewable energy
13	Theory of relativity	Nuclear weapons, big bang, satellite positioning
14	Schrödinger equation	Quantum physics, laser, computer chips
15	Information theory	Audio CD, Artificial Intelligence, Cryptography
16	May's Chaos Theory	Weather forecast, population dynamics, planetary motion
17	Black–Scholes equation	Financial markets, stock exchange

With meticulous scrutiny, I normalized the data and tallied the dates of their discovery. Over a span of 380 years, 16 of these equations were unveiled, a progression that Gumilev would categorize as inertial. Drawing upon the Mathematical Genealogy Project[147], I accounted for the multitude of mathematicians — 137,672 in number — and calculated their average representation in the population. Estimating roughly, I extended this distribution over the years, encompassing the entirety of our intellectual landscape, including realms such as economics and business. The result, a staggering figure of 8,800,000 individuals, emerged. Employing a simple calculation, I arrived at a coefficient of 0.000003 to the current population, or 0.0026 of all these potential passionaries, individuals with the capacity to transform into contextual passionaries. These remarkable beings would be entrusted with the task of pursuing endeavors vital for our well-being: combating cancer, addressing the ravages of time, fighting against environmental degradation, countering insidious propaganda, resisting the manipulation of public sentiment, advocating for advancements in teleportation, telepathy, long-range energy transmission, time travel, and deep space exploration[148]. The question lingers, will we discover 23,158 resolute and astute souls willing to dedicate themselves to these urgent imperatives? Will they rise to combat the plight of hunger, pollution, poverty, and the treachery perpetuated by our abhorrent governments,

---

[147] The Mathematics Genealogy Project.

[148] By the way, a breakthrough on the above issues in my understanding will require very little: namely, to concentrate on decoding the Riemann Zeta function in relation to biology, further generalization of the Dirac equation in relation to quantum physics, the study of photonic crystals from the point of view of the Fibonacci spiral, the study of Christoffel symbols for the general theory of gravitation, and, finally, the study of knowledge of the same reliability and level in any field to identify paired scientific Patterns.

transnational corporations, and criminal syndicates? Alas, I cannot discern the answer. Yet, were I among those individuals, I would not hesitate to stake my life on the line. Let there be no doubts, if the course is clear and the cause serves the betterment of our world, I would wholeheartedly embrace such sacrifice.

> Here are a few additional paths that can be pursued to bolster the relevance of the aforementioned code of honor and enhance humanity's prospects of endurance.

Education

Education holds a pivotal role in heightening awareness and advancing the values and tenets enshrined in the code of honor. Collaborative efforts between governments and civil society organizations must be undertaken to ensure that educational systems prioritize values like critical thinking, empathy, and emotional resilience.

Technological advancements

Technological progress can offer solutions to numerous challenges confronting humanity, be it climate change or resource depletion. Governments, corporations, and civil society organizations must invest in and advocate for technological innovations that align with the principles of the code of honor.

Global cooperation

The predicaments we face necessitate global solutions. Governments, international organizations, and civil society must forge alliances to address issues such as climate change, nuclear disarmament, and poverty. The principles articulated in the code of honor can provide a foundation for international collaboration and interaction.

Responsible consumption

Individuals also have a role to play in championing the principles of the code of honor through responsible consumption practices. This entails reducing waste, opting for eco-friendly products, and supporting ethical businesses.

Fortifying democratic institutions

Democratic institutions serve as platforms for citizen participation in decision-making processes that shape their lives. Governments must endeavor to strengthen these institutions, ensuring transparency, accountability, and inclusivity. This will help prevent the concentration of power and wealth in the hands of a privileged few.

By implementing these measures, we can reinforce the applicability of the code of honor and bolster humanity's chances of survival. Embracing the principles outlined in the code and working collectively to tackle the challenges confronting us can reduce the likelihood of the negative scenario envisioned earlier.

Various social, political, and legal instruments can be employed to augment the chances of applying the progressive principles. Socially, campaigns to raise awareness and educate the public can be organized, employing public lectures, seminars, and workshops to disseminate knowledge on the significance of environmental sustainability, peaceful coexistence, and respect for human rights.

Politically, governments can enact laws and policies that align with the code of honor. Such legislation can aim to curb greenhouse gas emissions, promote renewable energy sources, safeguard minority rights, and encourage the peaceful resolution of conflicts. Governments can establish institutions tasked with enforcing these laws.

Legally, international agreements and conventions can be ratified to ensure environmental sustainability, peace, and the protection of human rights. These agreements can serve as legal frameworks for resolving disputes, enforcing accountability, and holding individuals and corporations responsible for their actions. Notably, the United Nations has various conventions in place, such as the Paris Agreement on climate change, the Universal Declaration of Human Rights, and the Convention on the Prevention and Punishment of the Crime of Genocide.

Embracing these measures augments the prospects of the code of honor's applicability, thereby increasing humanity's chances of survival. However, no organization, institution, or convention can supplant our individual awareness, which lies at the core. Without it, all these endeavors become mere bureaucratic structures and hollow rhetoric. Each one of us must decide to thwart evil within ourselves and let in the Light, even if it entails temporary or, as the noblest among us would do, permanent material hardship. Failure to do so will soon witness our biological species fading away.

And here's what gives me confidence, among other things: when we multiply the height of the Great Pyramid of Giza[149] by 43,200, we get the polar radius of the Earth, and when we multiply the perimeter of its base by 43,200, we get the equatorial circumference of the Earth.

The number 43,200 is no mere chance. It corresponds to 600 cycles of a 1-degree precession of the Earth, which occurs every 72 years. I interpret this as a message from the architect of the game in which we are participants: four

---

[149] *Pyramid of Cheops* (Arabic. هرم خوفو), the Great Pyramid of Giza — the largest of the Egyptian pyramids, a monument of architectural art of Ancient Egypt; the first and only of the "Seven Wonders of the World" that has survived to this day, and the oldest of them: its age is estimated at about 4,500 years.

thousand years after the construction of this monument, some still believed the Earth was flat and rested on the backs of three elephants.

As the reader may have gathered, I firmly believe that the entire history of the world's existence is governed by absolute determinism. On one hand, this could plunge us into despair, considering that, in my humble opinion, we have about 200 billion years of existence in this universe. In the grand scope of the Universe's history, the survival or demise of a tiny flicker of intelligent life situated on Earth, within the Solar System, Orion Arm, Milky Way, Local Group, Virgo Supercluster, or even Laniakea Supercluster may hold significance. Or perhaps it may not. We cannot say for certain. However, if we assume that our thoughts create and recreate this world, that they give birth to the energy from which light emerges, and that this energy, through a decrease in vibration, brings forth elementary particles which, in turn, create matter, and when we interact with this matter, we generate emotions and formulate thoughts based upon them, thus continually recreating this world, then we become the primary source of energy fueling the expansion of the universe. We are responsible for the existence of all that resides within it.

Within the vast expanse of the universe, we occupy a minuscule planet on the periphery. Yet, it is our most cherished place in the universe — *Lalangamena*[150], our home. The vulnerability of our exquisite planet, primarily at the hands of our own species, fills me with unease.

On the other hand, it also fills me with joy and optimism, for I am now fully convinced of the divine plan for the existence of this world, where its Creator is myself, just as it is for each of us — Ronin of the Sun — those who have reached the conclusion of this book or those who

---

[150] See a beautiful story by. Gordon Dixon "Lalangamena".

require no books for such comprehension. Those who can gaze into the Creator of this world and its Adversary by merely looking into the mirror. When I contemplate this, my heart brims with love and pride for you and me, for those who venture into the black abyss with a smile upon their lips.

All decisions are made in the present moment, here and now. And it is precisely for such a decision, here and now, that the following pages offer a concise guide to awareness — the Code of Honor of the Ronin of the Sun.

I'm humbly on my knees, pressing my lips to the hands of those who have nearly reached the end of this book. Please, do turn to the next page.

The Code of Honor
of the Ronin of the Sun

is as follows:

1. I shall lead a mindful existence, cherishing my own happiness while honoring the well-being of others.

2. With honesty and bravery, I shall defend my ideals, moral values, family, and possessions.

3. I shall exhibit reverence for the boundaries of others, fostering kindness, love for nature, and the preservation of resources for future generations.

4. I shall champion education, self-improvement, and technological progress, while discouraging religious dogma, fascism, enslavement, manipulation, and undue enrichment.

5. I shall advocate for transparency in society, upholding democracy, humanistic principles, and the liberty of both myself and my fellow beings.

6. I shall embrace honesty in my familial and business relationships, with the understanding that certain boundaries exist for trade and family secrets.

7. I shall protect the vulnerable, lending support to those who stumble but strive to better themselves.

8. I shall demonstrate resourcefulness, sobriety, and respect for all individuals, including women, children, minorities, and those with predetermined destinies, such as people with disabilities.

9. I shall safeguard my society and ideals, utilizing resolute and equitable justice while considering the interests of others.

10. I solemnly pledge to uphold these enduring principles at any cost, even if it demands my life, and I shall be the sole arbiter of this decision.

ENDURING PRINCIPLES:

Happiness and Joy

The pursuit of happiness and joy shall reign supreme, while acknowledging the rights of others.

Respect

I shall honor the boundaries, interests, and resources of others and our planet.

Integrity

I shall exhibit honesty, courage, and transparency in all my endeavors and choices.

Education

I shall encourage learning, self-improvement, technological advancements, and the quest for knowledge.

Justice

I shall defend my society and ideals with resolute and unwavering justice.

Democracy

I shall support democracy, humanistic values, and the freedom of both myself and others.

Responsibility

I shall be resourceful, temperate, and accountable for my actions, while treating women, children, minorities, and those with predetermined destinies with respect.

Love

I shall foster kindness, love for nature, and the preservation of resources for future generations.

Oppression's Bane

I shall thwart religious dogma, fascism, slavery, manipulation, undue enrichment, and the concentration of power and wealth among a select few, and I myself shall not partake in such endeavors.

Halting the Spread of Evil

If unjust violence or any form of malevolence is inflicted upon me, I vow to contain the evil within myself and prevent its proliferation.

11. I believe that this code of honor and its enduring principles will guide those who strive for a conscientious existence, prioritizing personal happiness while respecting the well-being of others.
12. I recognize that freedom of choice is paramount to the state of joy, and I shall never impose the Code of Honor of the Ronin of the Sun upon others.

Pictures

Epilogue

I n the wake of remarkable and potentially highly dramatic events accompanying the active phase of the Earth's precession cycle, which has been discussed in this book, commonly referred to as the transition from the *Age of Pisces* to the *Age of Aquarius*, significant developments have already unfolded, with more tectonic changes on the horizon. The era shift will continue to bring about transformative shifts in the world over the course of several centuries. I, however, had the conscious privilege of catching a glimpse of the new era arrival early in my life. This was primarily due to my audacious belief in the existence of multiple life cycles and my inherent skepticism toward beliefs, sects, training systems, religions, political parties, and also owing to the guidance from my remarkable teachers in languages, karate, philosophy, history, mathematical statistics, law, religious studies, and more. Each mentor left their unique imprint, yet none of them fully captivated my attention.

The objectively tumultuous events of my life over the past decade led me through a series of internal decisions that have shaped a person entirely distinct from the one who initiated this transformative journey, one who has weathered countless trials, from the loss of communication with my son to the untimely demise of my parents, enduring unceasing attempts to bring me down and enduring constant peripetia[151] — changes of fortune.

In essence, about a third of the information of this book is coming from numerous individuals who became my teachers in Russia, Holland, Hong Kong, London, Ukraine, and the blessed United States of America, where

---

[151] *Twists and turns* (Greek: Νέμεσις). In modern language, it means a sudden unfavorable change of fate or an unexpected complication. In dramaturgy, it is a technique that denotes an unexpected turn in the development of the plot and complicates the plot. In his treatise Poetics, Aristotle defined peripeteia as "the transformation of action into its opposite."

I presently reside. I must express my gratitude to Lesya in Ukraine, who helped me write this book amid Russian bombs falling on Kharkiv, working late nights with darkened windows, as well as to Olivera in the United States, who taught me the unconditional love for oneself, the pure-hearted magician Olga, the wise Japanese Lily who taught me to discern temporary companions in life by observing their choice of vegetables in a grocery store, the witty American Bae, the magnificent master Takahashi who instructed me on transforming my body through the transformation of my thoughts, and my friends and partners across continents and time. I must also seize this moment to extend my gratitude to my adversaries: your hatred and attempts to harm me have made my life conscious. Each of these individuals has left an indelible mark on my existence. Their reflections are transmuted into this book, collectively forming an indivisible unity with me and, concurrently, serving as the fundamental raison d'être for my continued existence — a perpetual ascent beyond myself. According to the Kabbalistic tradition[152], only *two* reasons underpin death: *fulfilling one's destiny in this world* and *the inability to fulfill one's purpose in life.*

The writing of this book was not solely driven by my personal desire but stemmed from a clear directive

---

[152] *Kabbalahá* (Heb. קַבָּלָה — *"receiving, accepting; Tradition"*) is a religious-mystical, occult and esoteric movement in Talmudic Judaism, which appeared in the XII century and became widespread in the XVI century. Esoteric Kabbalah claims secret knowledge, divine revelation, encoded in the text of the Torah. Kabbalah is associated with the comprehension of the Creator and creation, the role and goals of the Creator, human nature, and the meaning of existence. The basis of Kabbalah is the ancient works "Yetzirah", "Baghir" and "Zo'ar". This doctrine is not a dogma, but I respect all the statistical conclusions that humanity has collected for millennia. Therefore, in my book, among others, I use the knowledge gained from this ancient and wise source for purely pragmatic purposes.

bestowed upon me directly from the source. I would like to conclude this passage on a beautiful note; however, for the sake of surgical precision, it must be acknowledged that two-thirds of this book is inscribed with the heartbeat, not with ink, but with the blood of an author who has made countless mistakes in life. This acknowledgment is necessary and thus, let us address it now.

This book emerges from the pulse, meaning that it comprises many recurring, akin segments that, in the language of cardiograms, might be termed cardiointervals. Though I cannot claim to have consciously chosen this style, as I wrote without contemplating the text's logical flow, a certain structure emerged after the book's completion. It involves elevating the reader to a figurative height — an avian flight — that allows an assessment of forthcoming trends. The reader can then apply this information to their daily life. Following this, I have presented the most crucial tools for enhancing the material aspects of life, after which we will endeavor to perceive an enlarged view of the world through the prism of the reader's immediate interests. Consequently, this book serves as a wholly pragmatic guide on how to live and attain happiness in the new era. Thestate of joy — above all else — is the sole value deemed worthy of divine attention by the author. Let us be unequivocal: the author does not deem anything other than happiness, including but not limited to concepts such as "duty," "homeland," or the bouquet of "social norms," as contextually significant and therefore does not lend them attention. The tools presented to the reader for study aim not merely to create the effects of delight and exhilaration, but to forge a material foundation enabling personal conviction to reasonably conclude that one is leading life correctly and adaptably to the external conditions at hand. The key to success in any endeavor lies in the ability to change one's mindset. I compelled myself to change by placing myself in situations where adaptation or outright failure were the only options. After countless failures, I began to transform. Though one failure would have sufficed, I suppose.

One more vital observation: this book serves as a survival guide amidst the swiftly changing conditions of everaccelerating time, rather than a book exclusively concerned with personal growth. Herein lie the ways to attain happiness and fulfillment despite the ground slipping from beneath our feet. Being content in these circumstances is akin to being alive. The sooner we comprehend this, the greater our likelihood of survival as not mere chance microorganisms, but as sentient beings.

Bernard Shaw imparted two additional paradigms to me: there is no point in seeking ourselves; we must create ourselves, and a rational individual adapts to the world rather than attempting to change it, whereas an irrational individual seeks to alter the world without first transforming themselves. Hence, all progress is driven by the irrational. While my irrationally stubborn nature does not inherently qualify me as someone capable of changing the world, and indeed, I am uncertain if I can do so, my ardent desire to effect change remains.

It is my fervent hope that readers will tear pages from this book, wrapping fish in them, writing in the margins, crossing out my ideas and replacing them with their own, leaving mocking annotations, scolding, and even laughing at its contents. Nevertheless, I implore them to read it through to the end, if only to criticize it with certainty, or consign it to waste paper, or erase it from memory moments after reading.

In the Japanese tradition, there exists a concept called shu ha ri: when the student is prepared, the teacher will appear, and when the student is truly ready, the teacher will disappear. Shu entails studying the fundamentals and emulating great mentors, ha involves experimentation and assimilating what has been learned from masters with personal experimentation, and ri signifies one's individual innovation and application in various situations.

Press forward! Time remains.

Content

© Oleksii Hrebeniuk,
design, layout, epub, 2023

ISBN: 979-821-824-076-9

Printed in USA

www.ingramcontent.com/pod-product-compliance
Lightning Source LLC
Chambersburg PA
CBHW062130040426
42335CB00039B/1891